D1045237

Contents

Preface

1. **An Emerging Paradigm for Research on Addition and Subtraction Skills**
Thomas A. Romberg 1

2. **The Development of Addition and Subtraction Problem-Solving Skills**
Thomas P. Carpenter and James M. Moser 9

An Analysis of Verbal Problems 10
Empirical Findings 13
General Discussion 20

3. **Levels of Description in the Analysis of Addition and Subtraction Word Problems**
Pearla Nesher 25

Regularities Found in Solving Word Problems of Addition and Subtraction 27
The Logical Structure 28
The Semantic Component 30
The Syntactic Component 35
Final Remarks 36

4. A Classification of Cognitive Tasks and Operations of Thought Involved in Addition and Subtraction Problems
Gérard Vergnaud 39

Basic Categories of Relationships 43
Experimental Results 48
Are Symbolic Representations Useful? 53
Conclusion 57

5. Interpretations of Number Operations and Symbolic Representations of Addition and Subtraction
J. Fred Weaver 60

Some Background Considerations 60
Some Instructional and Research Considerations 63

6. An Analysis of the Counting-On Solution Procedure in Addition
Karen C. Fuson 67

Counting All and Counting On 67
The Structure of the First Addend In the Counting-On Procedure 68
The Structure of the Second Addend in Counting On 73
Coordinating the First and Second Addends in Counting On 77
Conclusion 78

7. Children's Counting in Arithmetical Problem Solving
Leslie P. Steffe, Patrick W. Thompson and John Richards 83

Levels of Problem Solving 85
Summary 96

8. The Development of Addition and Subtraction Abilities Prior to Formal Schooling in Arithmetic
Prentice Starkey and Rochel Gelman 99

Some Early Competencies 99
Piagetian Theory 109
An Alternative View 112

9. Towards a Generative Theory of "Bugs"
John Seely Brown and Kurt VanLehn 117

The Form of Generative Theory 122
Repair Generation 125
Critics 131
Concluding Remarks 133

10. **Syntax and Semantics in Learning to Subtract**
Lauren B. Resnick 136

Distinguishing Syntax and Semantics 137
A Closer Look at Children's Semantic and Syntactic Knowledge 142
Linking Syntax and Semantics 148
Why Mapping Works 150

11. **General Developmental Influences on the Acquisition of Elementary Concepts and Algorithims in Arithmetic**
Robbie Case 156

Cross-Domain Parallels In Cognitive Development 156
The Role of Central Processing Capacity 160
Parallel Trends in the Area of Mathematics 162
Implications for Instruction 165

12. **The Structure of Learned Outcomes: A Refocusing for Mathematics Learning**
Kevin Collis 171

Background 171
The Response Model in Relationship to Addition and Subtraction 177
Conclusion 181

13. **Type 1 Theories and Type 2 Theories in Relationship to Mathematical Learning**
Richard R. Skemp 183

A New Model of Intelligence 183
A Refocusing 188

14. **The Development of Addition in Contexts of Culture, Social Class, and Race**
Herbert P. Ginsburg 191

Study I: Mental Addition in Cross-Cultural Context 193
Study II: Written Addition in Cross-Cultural Context 198
Study III: Social Class and Race in America 204
Conclusion and Implications 208

15. Learning to Add and Subtract: A Japanese Perspective
Giyoo Hatano 211

Development of Number Concept and Calculation Skills up to Kindergarten 211
Acquisition of Addition-Subtraction Skills During Elementary Grades 214
Does Japanese Culture Favor the Development of Calculation Skills? 217
Concluding Remarks 221

16. The Psychological Characteristics of the Formation of Elementary Mathematical Operations in Children
V. V. Davydov 224

The Origins of the Number Concept 225
The Basic Concept of Quantity 228
Curricular Implications 229
Evaluation 235
Conclusion 236

Author Index 239

Subject Index 243

Preface

A hallmark of much of the recent research on children's thinking has been the focus on explicit content domains. Much of this research has been represented by an eclectic collection of studies sampled from a variety of disciplines and content areas. However, in the last few years, research in several content domains has begun to coalesce into a coherent body of knowledge. The chapters in this work represent one of the first attempts to bring together the perspectives of a variety of different researchers investigating a specific, well defined content domain.

This book presents theoretical views and research findings of a group of international scholars who are investigating the early acquisition of addition and subtraction skills by young children. Together, the contributors bring a blend of psychology, educational psychology, and mathematics education to this topic. Fields of interest such as information processing, artificial intelligence, early childhood, and classroom teaching and learning are included in this blend.

Following a brief introductory chapter, the book is separated into five parts. The first part, "The Structure of Addition and Subtraction Problems," presents four different, but complementary, approaches to understanding how children think and operate on verbal addition and subtraction problems by describing structural features of verbal problems and classifying different problem types by those features. Using empirical data, the authors demonstrate interesting differences in how children solve each of the various problems. These essays raise significant questions about the relationship between the semantic structure of various problem types, the strategies children actually use in solving such problems, the use of symbolic representations, and the utility of symbolic instruction.

The second part is entitled "The Role of Counting in Solution Processes." These essays describe current research on counting and how young children solve many addition and subtraction problems using counting procedures.

In the third part, "The Analysis of Error Patterns," two related papers examine the errors children make on the common "fair-trading" subtraction algorithm from an information processing perspective. These papers provide an explanation of how children's invention, apparent in their counting solutions to simple problems, can lead to learning incorrect procedures to solve problems requiring algorithmic solutions.

The fourth part is entitled "Alternate Theoretic Perspectives." Three essays argue for theoretical refocusing of research based on an information processing orientation for studies of children's arithmetical development, an examination of the structure of children's learned outcomes, and a consideration of levels of cognitive decision making.

The essays in the concluding part, "The Development of Addition and Subtraction Skills in Other Cultures," focus on the cultural and ideological context within in which addition and subtraction skills develop. Data from Africa and America suggest there are common strategies and patterns across cultures and classes. However, there are linguistic and cultural factors in the Japanese culture that contribute to a seemingly different developmental pattern.

The chapters are an outgrowth of a seminar held in November 1979 at the Wingspread Conference Center in Racine, Wisconsin, a facility which encouraged sustained discussion of the ideas presented. The first drafts of essays were shared prior to the seminar.

Three of the essays prepared for the book were substantially longer and more developed papers which were presented at the conference and have been subsequently published as separate projects papers by the Wisconsin Research and Development Center.

Collis, K. *Cognitive development, mathematics learning, information processing and a refocusing.* (Project Paper No. 81-1) Madison: Wisconsin Research and Development Center for Individualized Schooling, 1981.

Skemp, R. R. *Theories and methodologies.* (Project Paper No. 81-3) Madison: Wisconsin Research and Development Center for Individualized Schooling, 1981.

Weaver, J. F. *Addition, subtraction, and mathematical operations.* (Project Paper No. 79-7) Madison: Wisconsin Research and Development Center for Individualized Schooling, 1979.

A number of other papers were significantly rewritten for publication, and ideas presented in the original papers are not always included in the published versions. Karen Fuson's paper underwent substantial revision. Copies of her original paper and an expanded version of the chapter prepared for this book are available from the author. In addition, V. V. Davydov, after reading the papers on counting skills, brought with him an English translation of a recently published paper of his on that subject and shared it with the participants. Because of its inaccessibility in English, it too has been published as a project paper.

Davydov, V. V., & Andronov, V. P. *The psychological conditions for the origination of ideal actions*. (Project Paper No. 81-2) Madison: Wisconsin Research and Development Center for Individualized Schooling, 1981.

Copies of these four project papers are available from the ERIC Center.

The seminar was co-sponsored by the Wisconsin Research and Development Center for Individualized Schooling, National Institute of Education, and The Johnson Foundation of Racine. In addition to the editors and authors, a number of scholars participated in the seminar at Wingspread and helped to clarify ideas and issues in the problems discussed in the book. They are:

Constance Martin Anick
University of Wisconsin-Madison

Arthur J. Baroody
Keuka College

Glendon Blume
University of Iowa

Anne Buchanan
University of Wisconsin-Madison

Connie Seaman Cookson
University of Wisconsin-Madison

Gabriela Delgado
Universidad Nacional Autonoma de Mexico

Ed Esty
National Institute of Education

Jane Donnelly Gawronski
San Diego County Public Schools

James G. Greeno
University of Pittsburgh

James Hiebert
University of Kentucky

Joan I. Heller
University of Pittsburgh

David Klahr
Carnegie Mellon University

Vicky L. Kouba
University of Wisconsin-Madison

Richard Lesh
Northwestern University

Mary Montgomery Lindquist
National College of Education

Douglas McLeod
National Science Foundation

Gary G. Price
University of Wisconsin-Madison

Richard J. Shavelson
University of California at Los Angeles

Merl C. Wittrock
University of California at Los Angeles

We are indebted to them for the discussion, constructive comments, and conversation.

We particularly want to thank Mr. Henry Halstead and his staff at Wingspread for their gracious hospitality. Not only are the facilities ideal for a seminar, but the care and cordiality of the staff helped create the collegial atmosphere which prevailed.

We would also like to thank Constance Martin Anick for her skillful handling of many of the administrative details of the Conference.

Finally, in preparing the book, we would like to thank each author for their patience and willingness to react to our suggestions. We wish also to thank Mary Pulliam for her care in editing the manuscripts and checking references, and Louise Smalley and Dorothy Egener for typing and retyping each of the chapters.

Thomas P. Carpenter
James M. Moser
Thomas A. Romberg

1 An Emerging Paradigm for Research on Addition and Subtraction Skills

Thomas A. Romberg
University of Wisconsin-Madison

For several centuries, being able to find "one's sums and differences" has been considered one mark of a schooled person. Although today we may have expanded our expectations about what constitutes literacy, we still expect all children to efficiently carry out operations on whole numbers. Yet, in spite of these expectations about the skills of addition and subtraction, there has been little consensus about how such skills develop. Lack of consensus does not mean there has been little research. Recent reviews (Carpenter, Blume, Hiebert, Anick, & Pimm, 1981; Carpenter & Moser, in press; Riley, Greeno, & Heller, in press) have identified an extensive body of research on addition and subtraction. Some of these studies have been quite influential. For example, Thorndike's instructional suggestions in his *Psychology of Arithmetic* (1922) became the model of how to teach arithmetic for decades,[1] and Brownell's (1947) research on subtraction demonstrated the superiority of the "decompositions" subtraction algorithm over the "equal additions" algorithm when taught with rational explanation. This made the "fair trading" procedure central to contemporary instruction in subtraction. But, to a large extent, the many studies on addition and subtraction represent an eclectic morass. This copious literature has lacked an implicit body of intertwined theoretical and methodological beliefs that permit selection, evaluation, and criticism. However, today we believe a change is imminent. The research and theoretical positions set forth in this volume should be viewed as foreshadowing the emergence of a firm research consensus in this area.

[1]Chapter 3 in Cronbach and Suppes (1969) presents a convincing argument about Thorndike's influence on mathematics instruction.

To build this argument, I follow Thomas Kuhn's description of the "route of normal science." In his now classic treatise on the growth of science, *The Structure of Scientific Revolutions* (1979), Kuhn argues that a significant turning point in the history of science occurs when from the chaos of competing ideas about a problem area, a single paradigm emerges which implicitly defines for practitioners the legitimate problems and methods of research. A paradigm gains that status because it is more successful than others in solving a few problems a group of researchers have recognized as acute. In this sense Kuhn argues paradigms have two essential characteristics. First, the paradigm's achievement in solving the acute problems is sufficiently unprecedented to attract a group of adherents. Simultaneously, the paradigm is open-ended, leaving all sorts of problems for the redefined group of practitioners to resolve. Kuhn calls the research carried out by this new group "normal science." It consists of actualizing the promise of the paradigm, "extending the knowledge of those facts that the paradigm displays as particularly revealing, by increasing the extent of the match between those facts and the paradigm's predictions and by further articulation of the paradigm itself" [p. 24].

It would be both presumptuous and incorrect to argue that a paradigm for research on the development of addition and subtraction skills has emerged and that the papers in this volume reflect work within a normal science. Rather, the papers reflect growing agreement around a constellation of ideas with the potential to become such a paradigm. Current work mirrors what Kuhn discusses as the "route of normal science."

Historically, the road to a research consensus in any area is arduous. In the absence of a paradigm all facts that could possibly pertain to the development of a given science are likely to seem equally relevant. As a result, early fact-gathering is a nearly random activity. Furthermore, in the absence of a reason for seeking some particular form of information, early fact-gathering is usually restricted to the wealth of data that lie ready at hand. Thus facts accessible to casual observation and experiment are pooled together with data retrievable from reports of classroom teaching, curriculum development, or evaluation.

Although this sort of fact-collecting has been essential to the origin of many significant sciences, one somehow hesitates to call the resulting literature scientific. Similarly, it would be hard to describe early studies on addition and subtraction as scientific (Carpenter, et al., 1981). This is true because such studies juxtapose facts that will later prove revealing with others that will for some time remain too complex to be integrated with theory at all. In addition, since any descriptions must be partial, such a typical natural history often omits from its immensely circumstantial accounts just those details that will be sources of important illumination to later scientists. Because the casual fact-gatherer seldom possesses the time or the tools to be critical, natural histories often relate reasonable descriptions with others that we are now quite unable to confirm. This is the situation that creates the intellectual morass characterizing the early stages of a science's development, and as Kuhn (1979) states:

> No wonder, then, that in the early stages of the development of any science different men confronting the same range of phenomena, but not usually all the same particular phenomena, describe and interpret them in different ways [p. 16].

With regard to the development of addition and subtraction skills, a set of scholars is confronting the same range of phenomena from essentially similar perspectives, and is beginning to reach a consensus on the acute problems to be solved, and beginning to use the same language and research methods to attack these problems. The emerging general paradigm is to formulate precise models of the cognitive processes used by subjects when carrying out specific tasks and how those processes change over time.

Brown (1970) argues the origins of this paradigm stem from two primary sources–computer simulation of cognitive processes and the writings of Jean Piaget. The basic strategy for this simulation of human processing was sketched by Herbert Simon (1962).

> If we can construct an information processing system with rules of behavior that lead it to behave like the dynamic system we are trying to describe, then this system is a theory of the child at one stage of the development. Having described a particular stage by a program, we would then face the task of discovering what additional information processing mechanisms are needed to simulate developmental change–the transition from one stage to the next. That is, we would need to discover how the system could modify its own structure. Thus, the theory would have two parts–a program to describe performance at a particular stage and a learning program governing the transitions from stage to stage [pp. 154–155].

In order to specify rules of behavior and modifications of behavior it is necessary to characterize the child as an organism functioning under the control of a developing set of central processes. Some of Piaget's notions of child development, such as schema, assimilation, and accommodation, have gradually become the basis for creating dynamic models of children's cognitive processes in solving specific problems. The rapprochement between these two quite different conceptualizations has not been easy, as Klahr and Wallace (1976) have argued.[2] Yet, today agreement on some aspects is emerging. The developing paradigm has four elements upon which there is some consensus:

1. detailed descriptions of the contexts within which specific tasks are embedded;
2. analyses of all the behaviors associated with the subjects' responses to performing the task;
3. repeated assessment of performance behaviors over time; and

[2]Klahr and Wallace (1976) in the preface to their examination of cognitive development from an information processing view.

4. inferences about the cognitive processing mechanism which relates information about the task with performance, and about changes in this performance.

For several reasons, children's processing of addition and subtraction information–the topic of this book–is one area where this emerging paradigm has proved revealing. Addition and subtraction are the first set of mathematical ideas typically taught in schools. Children bring to such problems well developed counting procedures, some knowledge of numbers, and some understanding of physical operations, such as "joining" and "separating," on sets of objects. Thus, from this context researchers have a unique opportunity to examine variations in how children process information prior to, during, and after formal instruction. Identifying stages of development in strategies children use to solve such problems is the basic problem addressed in this book.

To solve a typical problem one first must understand its implied semantic meaning. Quantifying the elements of the problem comes next (e.g., choosing a unit and counting how many). Then, the implied semantics of the problem must be expressed in the syntax of addition and subtraction. Next the child must be able to carry out the procedural (algorithmic) steps of adding and subtracting. Finally, the results of these operations must be expressed.

As a group, the papers in this volume employ a variety of descriptions for the various cognitive processes or subprocesses children use on such problems. As yet, there is no agreement on terms for describing the problem contexts, the types of processes, or the processing mechanisms children use. Nevertheless, there is agreement that our aim is to formulate precise models that describe children's addition and subtraction skills and how those skills change over time.

The importance of specifying task context is reflected in the chapters by Thomas Carpenter and James Moser, Pearla Nesher, and Gérard Vergnaud in this volume. Because addition and subtraction sentences can be used to represent a wide variety of problems with different semantic structures, it is important for these authors to classify different types of verbal problems, and to study whether children can solve such problems prior to formal instruction. If children can, investigation of whether they use different strategies with problems having different semantic structures, and investigation of the changes in choice of strategies, is appropriate. Thus, the notions of verbal comprehension and the strategies used to quantify, represent, and calculate are acute problems of interest.

In this volume J. Fred Weaver and Vasily Davydov present arguments about the conceptualization of problem context from a mathematical perspective. Weaver stresses an alternative "unary operation" notion about addition and subtraction whereas Davydov embeds addition and subtraction in a broader mathematical perspective which stresses quantification processes before operational processes.

With few exceptions the authors in this book go well beyond tallying the number of correct and incorrect responses when describing children's behaviors in response to addition and subtraction problems. Identification of the actions and strategies children use on specific tasks is central to the papers by Carpenter and Moser; Vergnaud; Leslie Steffe, Patrick Thompson, and John Richards; Karen Fuson; and Prentice Starkey and Rochelle Gelman. Examining errors for prevalent patterns is a major emphasis in the investigation of both John Seely Brown and Kurt Van Lehn, and Lauren Resnick.

Inferences about cognitive processes used to produce responses and changes in responses over time are based on simulation models in both Brown and Van Lehn's model and Resnick's research; on notions of developmental stages in the work of Carpenter and Moser; Starkey and Gelman; Steffe et al.; and on instruction in Nesher's research. Cultural background and its influence on performance is stressed in both Giyoo Hatano's research and in Herbert Ginsburg's studies. It should be noted that the latter two authors are on opposite sides of the issue. Hatano argues that cultural background is important and Ginsburg cites evidence that it is not. Finally, in three broader theoretical papers, Robbie Case, Kevin Collis, and Richard Skemp stress different considerations for future models of cognitive processing. Case emphasizes children's developing memory capacity and how information is organized for storage, Collis agrees with Case but stresses learned outcomes, and Skemp argues for a theoretical formulation positing a "director system" at two levels.

All the papers build models to explain children's behaviors. For example, because children bring to typical verbal problems well developed counting skills and use those skills to quantify and often solve such problems, the study of the development of counting skills themselves is of particular interest. Steffe et al. and Fuson examine this topic.

Carpenter and Moser, Vergnaud, Nesher, and Starkey and Gelman examine the way children represent or use representations of various problems. The use of physical manipulatives, pictorial illustrations, and symbolic statements is modeled by this group of researchers.

As previously argued, one feature of an emerging consensus on a paradigm is agreement on methods of inquiry into the questions of critical importance. In the research presented in this volume, clinical interviewing of students is the predominant methodology. Carefully designed tasks and probing questions presented to a small sample of children are accepted as appropriate procedures. The papers by Carpenter and Moser, Resnick, Steffe et al, and Fuson, in particular, reflect this strategy. Davydov, Steffe et al., and Resnick use the "teaching experiment" extension of this procedure.

In most chapters, the data are generally presented descriptively with little use of statistics to bolster the arguments. In fact, because the concern is on formulation of models, attention is drawn to questions which may not be answerable with usual methods of statistical inference.

Underlying the book is a belief that by using this paradigm, we will eventually derive information that can be used to improve instruction. In particular, Case and Davydov draw inferences for teachers based on current knowledge.

The strength of this emerging scholarly consensus on how addition and subtraction skills develop lies in the fact that new research can begin where the last left off. Such research can concentrate on subtle or even esoteric aspects of the phenomena, assured that findings will add new information to a conceptual whole.

The weakness of consensus on any paradigm rests primarily in the fact that adherence to a single perspective makes questions appear insignificant which are deemed critical from other perspectives. For example, the question, "Who decides what subtraction algorithm should be taught?" is critical to the curriculum theorist interested in the structure of the content to be covered. For a behavioral psychologist, answering the question "What extrinsic motivational procedures are effective in getting children to add or subtract with low error rates?" is critical. For the instructional designer, it may be critical to answer the question "Which of two (or more) sequences of instruction is more efficient and effective?"

Since no perspective is all-encompassing, choosing a paradigm limits the variety of "acute" questions. In this book, the choice of a cognitive perspective limits the important questions to how children construct meanings for addition and subtraction situations and how those meanings change over time.

During the past decade, the authors of these chapters have carried out a great deal of significant work which is now coming together. This volume clearly reflects the emergence of a "normal science" approach to studying the development of addition and subtraction skills. What is important is to appreciate the growing consensus on the phenomena of interest, the acute problems to be studied, and the appropriate research methodology being used.

REFERENCES

Brown, R. Introduction. In Society for Research in Child Development (Ed.), *Cognitive development in children.* Chicago: University of Chicago Press, 1970.

Brownell, W. An experiment on "borrowing" in third-grade arithmetic. *Journal of Educational Research,* 1947, *41* (3), 161–171.

Carpenter, T., Blume, G., Hiebert, J., Anick, C., & Pimm, D. *The development of addition and subtraction concepts: A review.* (Theoretical Paper) Madison: Wisconsin Research and Development Center for Individualized Schooling, 1981.

Carpenter, T., & Moser, J. The acquisition of addition and subtraction concepts. In R. Lesh & M. Landau (Eds.), *Acquisition of mathematics on concepts and processes.* New York: Academic Press, in press.

Cronbach, L., & Suppes, P. (Eds.). *Research for tomorrow's schools: Disciplined inquiry for education.* New York: The Macmillan Company, 1979.

Klahr, D., & Wallace, J. *Cognitive development: An information-processing view.* Hillsdale, NJ: Lawrence Erlbaum Associates, Inc., Publishers, 1976.

Kuhn, T. *The structure of scientific revolutions* (2nd edition). Chicago: The University of Chicago Press, 1979.

Riley, M., Greeno, J., & Heller, J. The development of children's problem solving ability in arithmetic. In H. Ginsburg (Ed.), *The development of mathematical thinking*. New York: Academic Press, in press.

Simon, H. An information processing theory of intellectual development. *Monographs of the Society for Research in Child Development,* 1962, *27*(2, Serial No. 82).

Thorndike, E. *The psychology of arithmetic*. New York: The Macmillan Company, 1922.

2 The Development of Addition and Subtraction Problem-Solving Skills

Thomas P. Carpenter
James M. Moser
University of Wisconsin-Madison

A tacit assumption of most school mathematics programs is that addition and subtraction are best introduced through physical or pictorial representations of joining or separating sets of objects. Another common assumption is that verbal problems are difficult for children of all ages, and children must master addition and subtraction operations before they can solve even simple verbal problems. A growing body of research indicates that both assumptions may be false. The results presented in this chapter indicate that before children receive formal instruction in addition and subtraction, many of them can successfully solve basic addition and subtraction word problems. This suggests that verbal problems may give meaning to addition and subtraction and in this way could represent a viable alternative for developing addition and subtraction concepts in school.

Our interest is in the word problems commonly found in elementary mathematics textbooks, which can be solved by a single operation of addition or subtraction. This is not to suggest that children necessarily solve these problems by adding or subtracting. In fact, young children generally do not solve them by applying an arithmetic operation.

A substantial body of research indicates that young children solve addition and subtraction computation exercises by using several basic counting strategies (Groen & Parkman, 1972; Groen & Resnick, 1977; Suppes & Groen, 1967; Woods, Resnick, & Groen, 1975). The same basic strategies are used to solve simple word problems (Carpenter, Hiebert, & Moser, 1981; Carpenter & Moser, 1979; Gibb, 1956; Steffe, Spikes, & Hirstein, 1976). However, because a variety of semantically different word problems can be solved by addition or subtraction, the choice of strategy becomes somewhat more complex.

In this chapter, we describe the strategies children use to solve addition and subtraction word problems before they receive formal instruction in addition and subtraction, and how these strategies evolve during the first year of instruction. Children's strategies are strongly influenced by the semantic structure of the problem situation. Therefore, first it is necessary to characterize major differences between different addition and subtraction problems.

AN ANALYSIS OF VERBAL PROBLEMS

Previous research has taken several approaches to characterize verbal problems. One is to classify problems in terms of syntax, vocabulary level, number of words in a problem, etc. (Jerman, 1973; Suppes, Loftus, & Jerman, 1969). A second approach differentiates among problems in terms of the open sentences they represent (Grouws, 1972; Lindvall & Ibarra, 1980; Rosenthal & Resnick, 1974). We have chosen a third alternative that considers the semantic characteristics of the problem. Our analysis is generally consistent with other analyses based on problem structure (Gibb, 1956; Greeno, 1978; Nesher & Katriel, 1978; Vergnaud, this volume), although we distinguish among problem types somewhat differently.

We have identified several basic dimensions that characterize the actions or relationships involved in addition and subtraction word problems. The first dimension is based on whether an active or static relationship between sets or objects is implied in the problem. Some problems contain an explicit reference to a completed or contemplated action causing a change in the size of a quantity given in the problem. In other problems no action is implied; that is, there is a static relationship between quantities given in the problem.

The second dimension involves a set inclusion or set–subset relationship. In certain problems, two of the entities involved are necessarily a subset of the third. In other words, either the unknown quantity is made up of the two given quantities, or one of the given quantities is made up of the other given quantity and the unknown. In other situations one of the quantities involved in the problem is disjoint from the other two. In this case a comparison of the two disjoint quantities is implied.

For problems that involve action, there is a third dimension. The action described in a problem may result in an increase or decrease in the initial given quantity. Because the static problems do not involve changing the given quantities, this dimension does not apply to them. Altogether there are a total of six different classes of problems based on these distinctions. We have labeled these six classes Joining, Separating, Part–Part–Whole, Comparison, Equalizing–Add On, and Equalizing–Take Away.

Joining, Separating, and Equalizing problems all involve action, whereas Part–Part–Whole and Comparison problems describe static relationships between quantities. Equalizing problems are distinguished from Joining and Separating problems on the basis of set–subset relationships. A similar distinction differentiates Comparison and Part–Part–Whole problems. In other words, for Joining, Separating, and Part–Part–Whole problems two of the quantities are a subset of the third. Equalizing and Comparison problems involve comparing disjoint sets. The distinction between Joining and Separating problems and between the two Equalizing problems is based on whether the described action is an increase or a decrease. Joining and Equalizing–Add On involve an increase; Separating and Equalizing–Take Away involve a decrease.

Basically, Joining is the process of actively putting together two quantities. The problems generally give an initial quantity and a direct or implied action that causes an increase in that quantity. Separating problems have the same characteristics as Joining problems except that the action involves a decrease. In Separating problems a subset is removed from a given set. Part–Part–Whole problems describe a static relationship between an entity and its two parts. Problems in the Comparison class involve comparing two disjoint quantities. This includes problems in which the difference between two quantities is to be found as well as problems in which one of two quantities and the difference between them are given and the second quantity is the unknown. Equalizing problems involve the same sort of action that is found in Joining and Separating problems, but there is also a comparison involved. Basically, Equalizing involves changing one of two entities so that the two are equal on some attribute. Equalizing–Add On involves an increase in the smaller quantity; Equalizing–Take Away involves a decrease in the larger quantity.

This classification scheme characterizes the types of action or relationships that are represented by most addition and subtraction problems. However, a fourth variable must be taken into account to completely characterize addition and subtraction problems: the nature of the unknown. For each of the six classes of problems, there are as many as three distinct problem types, depending on which quantities are given and which is the unknown. Although the action or relationship involved in each class of problems is essentially the same, the problem types are very different and potentially involve different methods of solution. In fact, each of the six basic classes or problems includes both addition and subtraction problems. Furthermore, between the possible problems in a class there are significant differences in difficulty that are a function of which quantities are given and which is the unknown (Grouws, 1972; Lindvall & Ibarra, 1980). Examples of each of the 17 distinct problems generated by this scheme are presented in Table 2.1.

Limitations. The framework that we propose to characterize addition and subtraction word problems is limited to simple problems that are appropriate for primary age children. It is not as complete as the framework proposed by Vergnaud in this volume that extends to operations on integers. Although our framework does not unambiguously characterize all addition and subtraction word problems, it has been useful to help us clarify distinctions between problem

TABLE 2.1
Verbal Problem Types

1. Joining	Connie had 5 marbles. Jim gave her 8 more marbles. How many marbles did Connie have altogether? Connie has 5 marbles. How many more marbles does she need to have 13 marbles altogether? Connie had some marbles. She won 8 more marbles. Now she has 13 marbles. How many marbles did Connie have to start with?
2. Separating	Fred had 11 pieces of candy. He gave 7 pieces to Linda. How many pieces of candy did Fred have left? Fred had 11 pieces of candy. He lost some of the pieces. Now he has 4 pieces of candy. How many pieces of candy did Fred lose? Fred has some candy. He gave 7 pieces to Linda. Now he has 4 pieces left. How many pieces of candy did Fred have to begin with?
3. Part–Part–Whole	There are 6 boys and 8 girls on the soccer team. How many children are on the team altogether? Brian has 14 flowers. Eight of them are red and the rest are yellow. How many yellow flowers does Brian have?
4. Comparison	There are 6 boys and 8 girls on the soccer team. How many more girls than boys are there on the team? Luis has 6 pet fish. Carla has 2 more fish than Luis. How many fish does Carla have? Luis has 6 fish. This is 2 more fish than Carla has. How many fish does Carla have?
5. Equalizing–Add On	There are 6 boys and 8 girls on the soccer team. How many boys should join the team so that there will be the same number of boys as girls on the team? There were 6 boys on the soccer team. Two more boys joined the team. Now there is the same number of boys as girls on the team. How many girls are on the team?

(*continued*)

TABLE 2.1
(*Continued*)

	Eight guests came to dinner. I added 2 more places at the table so there would be the same number of place settings as guests. How many place settings did I have to begin with?
6. Equalizing–Take Away	There are 7 cups and 11 saucers on the table. How many saucers should I put away to have the same number of cups as saucers?
	There were 11 glasses on the table. I put 4 of them away so there would be the same number of glasses as plates on the table. How many plates were on the table?
	There were some girls in the dancing group. Four of them sat down so that each boy would have a partner. There are 7 boys in the dancing group. How many girls are in the dancing group?

types and to help us distinguish between problems with clearly different semantic characteristics as opposed to problems that merely differ in terminology or syntax. Furthermore, the results discussed in the next section of this chapter indicate that children's solution processes clearly reflect our distinctions between problem types.

EMPIRICAL FINDINGS

The relationship between the structure of addition and subtraction problems and the processes that children use to solve them has been the focus of a 3-year longitudinal study begun in fall 1978. The following data from the first year of the study describe the problem-solving strategies of 150 first-grade children. These results and the conclusions that follow from them are supported by several related studies (Blume, 1981; Carpenter et al., 1981; Carpenter, Moser, & Hiebert, 1981; Hiebert, 1981) and are consistent with the findings of other recent studies of addition and subtraction word problems (Gibb, 1956; Hebbeler, 1977; Lindvall & Ibarra, 1980; Nesher & Katriel, 1978; Riley, 1979).

In the first year of our longitudinal study, each child was individually interviewed in September, January, and May. At the time of the first two interviews, no formal instruction on addition or subtraction had occurred. At the time of the third interview, the children had received 2 months of instruction on addition and subtraction for sums between 0 and 10. Details of the study can be found in Carpenter and Moser (1979).

Addition

The interviews included Joining and Part–Part–Whole addition problems. Each problem type was presented under four different conditions resulting from the crossing of two variables, number size, and the availability of manipulative aids. The manipulative dimension involved the presence or absence of physical objects that could be used to represent the action or relationships described in the problems. Number size included a set of smaller number triples, the sum of whose addends was between 5 and 9, and a larger set for which the sum was between 11 and 16. The following problems are representative of the problems administered:

Joining	Wally had 3 pennies. His father gave him 5 more pennies. How many pennies did Wally have altogether?
Part–Part–Whole	Sara has 6 sugar donuts. She also has 9 plain donuts. How many donuts does Sara have altogether?

Strategies. Three basic levels of addition strategies were identified: strategies based on direct modeling with fingers or physical objects; strategies based on the use of counting sequences; and strategies based on recalled number facts. In the most basic strategy, physical objects or fingers are used to represent each of the addends, and then the union of the two sets is counted starting with one. This strategy is called *Counting All With Models*. Theoretically, there are two ways in which this basic strategy might be carried out. Once the two sets have been constructed, they could be physically joined by moving them together or adding one set to the other, or the total could be counted without physically joining the sets. This distinction is important. The first case would best represent the action of the Joining problems, whereas the second would best represent the static relationships implied by the Part–Part–Whole problems. However, although we observed clear examples of each type of strategy, children generally did not distinguish between the two strategies in solving either Joining or Part–Part–Whole problems. It was often difficult to classify a strategy as clearly representing either of the two cases. Thus, we have concluded that there is a single Counting-All-With-Models strategy. The strategy may be accompanied by different ways of organizing the physical objects, but the arrangements do not represent distinct strategies or different interpretations of addition.

A third alternative is also possible. A child could construct a set representing one addend and then increment this set by the number of elements given by the other addend without ever constructing a second set. Such a strategy would seem to best represent a unary conception of addition (see Weaver, this volume). This strategy was almost never observed.

Three distinct strategies involving counting sequences were observed. In the most elementary strategy, the counting sequence began with one and continued until the answer was reached. This strategy, which is called *Counting All With-*

out Models, is the SUM strategy identified by Suppes and Groen (1967) and Groen and Parkman (1972). It is similar to the Counting-All-With Models strategy except that physical objects or fingers are not used to represent the addends. However, this strategy and the two following counting strategies require some method of keeping track of the number of counting steps that represent the second addend in order to know when to stop counting. Most children would simultaneously count on their fingers, but a substantial number gave no evidence of any physical action accompanying their counting. When counting was carried out mentally, it was difficult to determine how a child knew when to stop counting. Some children appeared to use some sort of rhythmic or cadence counting. Others explicitly described a double count, but children generally had difficulty explaining this process. When fingers were used, they appeared to play a very different role than in the direct-modeling strategy. In this case, the fingers did not seem to represent the second addened per se, but were used to keep track of the number of steps in the counting sequence. When using fingers, children often did not appear to have to count their fingers, but could immediately tell when they had included a given number of fingers.

The other two counting strategies are more efficient and imply a less mechanical application of counting. In applying these strategies, a child recognizes that it is not necessary to reconstruct the entire counting sequence. In *Counting On From First,* a child begins counting forward with the first addend in the problem. The *Counting-On-From-Larger* strategy is identical except that the child begins counting forward with the larger of the two addends. In our studies, this has always been the second addend. This strategy is the MIN strategy of Groen and Parkman (1972).

Although the children in the study had not been taught number facts until the latter part of the school year, some of them learned a great deal about addition outside of school, including a wide range of number facts. These children were sometimes able to apply their knowledge of addition facts to solve simple verbal problems. In some instances certain children would use a *Known Fact* involving the numbers given in the problem. In other instances some children used *Derived Facts* generated from a small set of known basic facts. Derived Facts usually were based on doubles or numbers whose sum is 10. For example, to solve a problem representing $6 + 8 = ?$, a child responded that $6 + 6 = 12$ and $6 + 8$ is just 2 more than 12. In an example involving the operation $4 + 7 = ?$, another child responded that $4 + 6 = 10$ and $4 + 7$ is just 1 more than 10.

Addition Results. At each interview, under all four problem conditions, the same pattern of responses was found for Joining and Part–Part–Whole problems. This suggests that there is very little difference in the way children approach these two types of problems. This is consistent with our observation that children do not distinguish between a strategy that includes a physical joining of two sets and one that does not.

Selected results for the Part–Part–Whole problems are summarized in Table

2.2. Although Interviews 1 and 2 were conducted before children received formal instruction in addition, most children were able to solve the simplest addition problems.

Some children appear able to represent and solve problems involving small numbers before they can solve similar problems involving larger numbers. In theory, the process of solving problems with small numbers or larger numbers is the same when physical objects are available. But the problems with smaller numbers were significantly easier.

The results suggest that the first strategy that children use to solve these two types of addition problems is the basic Counting-All-With-Models strategy. This strategy was used most frequently in the first two interviews. There was a substantial decline in its use with small numbers in the third interview. In the third interview, there was a corresponding increase in the use of Counting-On strategies, especially with larger numbers. There was very little evidence of children using a Counting All Without Models strategy.

Of particular interest was the fourth interview condition where larger numbers were used but no physical aids were available. Because it is more difficult to represent numbers larger than 10 with fingers, many children opted to use the Counting-On strategies rather than the less advanced Counting-All strategy they would use when physical aids were present. It appears that this condition forced many children who were content to use objects to model a problem when they were available to use the more advanced Counting-On strategies.

TABLE 2.2
Results for Part-Part-Whole Addition Problems

			STRATEGY (Percent Responding)					
			Counting All		*Counting On*		*Numerical*	
Condition	*Interview*	*Percentage Correct*	*With Models*	*Without Models*	*From First*	*From Larger*	*Recalled Fact*	*Derived Fact*
Smaller numbers Physical objects	1	75	54	5	5	6	6	3
	2	82	52	1	5	8	13	6
	3	91	33	1	4	19	30	1
Smaller numbers No objects	1	64	38	6	6	9	8	3
	2	72	31	3	9	9	18	6
	3	91	14	1	13	16	36	0
Larger numbers Physical objects	1	50	51	1	3	3	1	1
	2	71	49	0	10	13	2	3
	3	84	45	1	9	25	9	4
Larger numbers No objects	1	28	16	1	8	8	1	1
	2	45	17	2	8	17	3	4
	3	66	13	0	21	33	6	3

Subtraction

Subtraction problems were administered under the same conditions as the addition problems. The following problems are representative of the four types of subtraction problems included in the study:

Separating	Tim had 11 candies. He gave 7 candies to Martha. How many candies did Tim have left?
Part–Part–Whole	There are 6 children on the playground. Four are boys and the rest are girls. How many girls are on the playground?
Comparison	Joe has 3 balloons. His sister Connie has 5 balloons. How many more balloons does Connie have than Joe?
Joining	Kathy has 5 pencils. How many more pencils does she have to put with them so she has 7 pencils altogether?

Strategies. Each of the three levels of abstraction observed for addition strategies also was found in the solution of subtraction problems. However, whereas a single basic interpretation of addition was observed, three distinct classes of subtraction strategies were found at the direct modeling and counting levels.

One of the three basic strategies involves a subtractive action. In this case, the larger quantity in the subtraction problem is initially represented and the smaller quantity is subsequently removed from it. When concrete objects are used, the strategy is called *Separating From.* The child constructs the larger given set and then takes away or separates, one at a time, a number of objects equal to the given number in the problem. Counting the set of remaining objects yields the answer. There is also a parallel strategy based on counting called *Counting Down From.* A child initiates a backward counting sequence beginning with the given larger number. The backward counting sequence contains as many counting number words as the given smaller number. The last number uttered in the counting sequence is the answer.

The second pair of strategies involves an additive action. In an additive solution, the child starts with the smaller quantity and constructs the larger. With concrete objects (*Adding On*) the child sets out a number of cubes equal to the smaller given number (an addend). The child then adds cubes to that set one at a time until the new collection is equal to the larger given number. Counting the number of cubes added on gives the answer. In the parallel counting strategy (*Counting Up From Given*) a child initiates a forward counting strategy beginning with the smaller given number. The sequence ends with the larger given number. Again, by keeping track of the number of counting words uttered in the sequence, the child determines the answer.

The third basic strategy is called *Matching*. Matching is only feasible when concrete objects are available. The child puts out two sets of cubes, each set standing for one of the given numbers. The sets are then matched one-to-one. Counting the unmatched cubes gives the answer.

Children could apply these strategies in several ways. The different strategies could represent different children's concepts of subtraction. If this were the case, each child would use a single strategy to solve all subtraction problems, although different children may use different strategies.

An alternative hypothesis is that different strategies would be used by a single child, depending on the structure of the problem. As we have just seen, certain of the strategies naturally model the action described in specific problems. The Separating problem is most clearly modeled by the Separating and Counting-Down-From-Given strategies. On the other hand, the implied joining action of the Joining (missing addend) problems is most closely modeled by the Adding-On and Counting-Up strategies. Comparison problems deal with static relationships between sets rather than with action. In this case, the Matching strategy appears to provide the best model.

For the Part–Part–Whole subtraction problem, the situation is more ambiguous. Because Part–Part–Whole problems have no implied action, neither the Separating nor Adding-On strategies (or their counting analogues), which involve action, exactly model the given relationship between quantities. And because one of the given entities is a subset of the other, there are not two distinct sets that can be matched.

In the next section, we present evidence that children tend to model the action or relationship described in the problem rather than attempting to relate the problem to a single operation of subtraction. In other words, the second hypothesis best characterizes children's solution strategies.

Subtraction Results. The general pattern of results for the subtraction problems was consistent with many of those found for addition. Children were more successful with smaller numbers than with larger numbers; they started with concrete direct-modeling strategies and subsequently shifted to the more abstract counting strategies. However, unlike addition, children employed a number of basically different subtraction strategies. Selected subtraction results for the experimental condition involving the larger set of numbers with physical objects available are summarized in Table 2.3.

The results indicate that the dominant factor in determining children's strategy was the structure of the problem. The strategy used by the great majority of children modeled the action or relationship described in the problem. This was true through all three interviews and under all problem conditions. For the Separating problem, almost all children used a subtractive strategy (Separating or

TABLE 2.3
Selected Results for Subtraction Problems[a]

			Strategy (Percent Responding)						
			Subtractive		*Additive*			*Numerical*	
Problem	*Interviews*	*Percentage Correct*	*Separate*	*Count Down From*	*Add On*	*Count Up From Given*	*Match*	*Recalled Fact*	*Derived Fact*
	1	42	56	1	1	0	0	1	1
Separating	2	61	68	1	1	3	0	1	2
	3	74	64	9	1	6	0	4	6
	1	39	5	0	31	9	3	1	1
Joining	2	57	2	0	43	12	1	2	4
	3	76	3	0	47	18	2	8	6
	1	38	13	0	4	6	18	1	1
Comparison	2	41	8	0	3	9	30	1	1
	3	57	14	1	5	11	33	3	5
	1	35	38	1	5	2	0	0	2
Part–Part–Whole	2	45	45	0	4	3	0	2	2
	3	70	55	6	2	7	1	0	5

[a] Larger number pairs, physical objects available.

Counting Down From). For the Joining (missing addend) problem, almost all children used an additive strategy (Adding On or Counting Up From Given). The results were not quite as overwhelming for the Comparison problem, but the Matching strategy was the most frequently used strategy when physical objects were available. In general, this strategy is not possible when objects are not available to construct the two sets to put in one-to-one correspondence. When concrete objects were not available, the problem was significantly more difficult. In this case, children tended to solve it using an Adding-On or Counting-Up-From-Given strategy.

The ambiguity of the Part–Part–Whole problem is reflected in children's selection of strategies. Although a majority tended to use a subtractive strategy, the additive strategies were used by a significant minority, especially in the fourth condition where manipulative objects and sufficient fingers were not available to model the separating process.

Children's tendency to use a strategy that represented the structure of the problem not only occurred at the early direct-modeling stage, but appeared to continue as they began to shift to the more abstract counting strategies. Although at the time of the third interview most of the children in our sample continued to use a strategy that represented the action or relationship described in the problem, almost half of them were using the more abstract Counting-Up-From-Given strategy when physical objects were not available rather than the more concrete Adding-On strategy.

The Counting-Down-From strategy, however, was used infrequently. Although a subtractive strategy was almost universally used to solve the Separating problem, children tended to use the Separating strategy with physical objects or fingers. When concrete aids were not available, 43% of the children used the Counting-Up-From-Given strategy to solve the Joining (missing addend) problem, but only 12% used Counting Down From to solve the Separating problem. Counting Down is a difficult process. When explicitly asked to count backwards a given number of steps, only about 50% of the first-graders in our sample could do so. Although our data are not conclusive in this regard and others have identified Counting Down From as a basic subtraction strategy (Woods et al., 1975), we would conjecture that some children never use a Counting-Down-From strategy prior to learning basic subtraction facts.

GENERAL DISCUSSION

For the six types of addition and subtraction problems included in the longitudinal study, some general patterns of development are apparent. Although different levels of abstraction are involved in the strategies that children use to solve simple Joining and Part–Part–Whole word problems, it appears that children have a reasonably unified concept of addition. They start out with a single basic strategy that involves directly representing the quantities described in the problem using physical objects. They are first successful in modeling and solving problems with sums less than 10 and soon extend their competence to problems with sums between 10 and 20 as long as a sufficient number of physical objects are available to model the problem. During the first grade, many children develop Counting-On strategies. It is not clear whether they initially use Counting On From First and then shift to a more efficient Counting-On-From-Larger strategy or acquire both simultaneously. If Counting On From First develops first, the Counting-On-From-Larger strategy appears to emerge in a very short time. It seems that many children must first learn to count on before they can solve addition problems with sums greater than 10 when physical objects are not available.

Early in their development of subtraction concepts, children have a variety of strategies for solving different subtraction problems. There is a general overriding tendency to model the action or relationship described in a problem that manifests itself as children generate strategies that provide different interpretations of subtraction. We would hypothesize that at first, children do not recognize the interchangeability of these strategies. This would account for the close match observed between problem structure and strategy.

Woods et al. (1975) hypothesized that older children are able to choose the most efficient of the counting strategies to solve numerical problems. So far, we have no data to support this conclusion with regard to children's solutions of verbal problems. Our data suggest that younger children have several independent conceptions of subtraction. A completely developed single concept of subtraction should involve integration of all these interpretations. Apparently, however, the initial transition involves a shift from the concrete strategies that completely model the problem to more abstract counting strategies. In other words, the ability to choose between strategies representing different interpretations of subtraction seems to develop after the ability to use more abstract versions of a given strategy in a particular problem.

Although children do not use different strategies for different addition problems, the three dimensions that characterize the different action and relationships involved in addition and subtraction problems are directly related to the strategies that children use to solve subtraction problems. This is not surprising as the dimensions represent major semantic differences between problems, and the strategies that children use represent an attempt to model the semantic structure of the problem. The set-inclusion dimension separates problems for which Matching is a major strategy from those for which Matching is almost never used. In addition to the results reported in the foregoing, an early study by Carpenter et al. (1981) found Matching to be the primary solution strategy for Comparison and Equalizing problems, whereas it was infrequently used for the Joining, Separating, or Part–Part–Whole problems. Similarly, the increase–decrease dimension separates problems in which additive strategies are used from problems in which subtractive strategies are used.

The distinction between Joining and Part–Part–Whole problems is also consistent with this analysis. The Part–Part–Whole problems involve neither an increase or decrease; consequently, either Joining or Separating strategies might be used. The Part–Part–Whole addition problem generates the same additive solutions as the Joining addition problem, but the Separating strategy is frequently used for the Part–Part–Whole subtraction problems rather than the Adding-On strategy that is generally used for the Joining problem.

Although children were relatively successful with the problems included in the longitudinal study, all addition and subtraction problems are not so easy. In fact, although children seemed not to differentiate between Joining and Part–Part–Whole addition problems, it has been demonstrated that they are much less successful with other types of addition problems (Carpenter et al., 1981; Grouws, 1972; Lindvall & Ibarra, 1980). In the longitudinal study, we have been concerned with the processes that young children use to solve addition and subtraction problems. Consequently, we included problems that logical analysis or empirical evidence would suggest are most likely to be solved. In general, these problems are ones in which the action or relationships described in the problem can be directly modeled without trial and error. In other problems, like the following missing minuend problem, this is not the case:

> Mary had some marbles. After she lost 5 of them, she had 8 marbles left. How many marbles did Mary have to start with?

In this problem, the initial state is the unknown quantity. To directly model this action would require some sort of trial-and-error strategy in which children guess at the size of the initial set and check their guess by removing 5 elements to see if there are 8 elements left. It is possible that this sort of problem will generate trial and error variations of the strategies that we have identified. Rosenthal and Resnick (1974) also suggest that trial and error strategies might be employed for this type of problem. However, we are aware of very little empirical evidence that indicates children systematically use trial and error strategies to solve these problems rather than transforming them so they can be solved directly.

The analysis presented here might help explain differences in difficulty between different problem types. It is reasonable that problems in which the quantities given could be represented directly would be easier than problems in which trial and error was necessary. This would explain why Separating problems like the foregoing are significantly more difficult than related Separating or Missing-Addend problems.

The results discussed in this chapter focus on the strategies that children use prior to and during the first few months of formal instruction in addition and subtraction. Consequently, we have provided a glimpse of only the early stages in acquisition of addition and subtraction concepts. Although there is evidence that children continue to use counting well into their middle school years (Lankford, 1974), most older children eventually use number facts and algorithms to solve the simple word problems described in this chapter. We believe that the transition from the informal modeling and counting strategies that children invent to solve basic addition and subtraction problems to the use of the memorized number facts and formal algorithms they learn in school is a critical stage in children's learning of mathematics. In the interviews described in this chapter, children were able to analyze and represent the structure of different problems in order to figure out how to solve them, and they were able to invent a variety of relatively sophisticated strategies for solving problems for which they had no algorithm. Other research has clearly documented that by the age of 9, many children mechanically add, subtract, multiple, or divide whatever numbers are given in a problem with little regard for the problem's content (Carpenter, Corbitt, Kepner, Lindquist, & Reys, 1980). Somehow in learning formal arithmetic procedures, many children stop analyzing the problems they attempt to solve. We hypothesize that older children's difficulty in analyzing and solving problems can be traced in part to the transition from using informal problem-solving strategies to memorized number facts and formal algorithms.

ACKNOWLEDGMENTS

The authors wish to gratefully acknowledge the assistance and critical comments made to earlier drafts of this chapter by Connie Martin Anick, Glen Blume, Anne Buchanan, Connie Cookson, Joan Heller, and Vicky Kouba.

Published by the Wisconsin Research and Development Center for Individualized Schooling, the project presented or reported herein was performed pursuant to Center Grant No. OB–NIE–G–80–0117 from the National Institute of Education, Department of Health, Education, and Welfare. However, the opinions expressed herein do not necessarily reflect the position or policy of the National Institute of Education, and no official endorsement by the National Institute of Education should be inferred.

REFERENCES

Blume, G. L. *Kindergarten and first-grade children's strategies for solving addition and subtraction problems in abstract and verbal problem contexts.* Unpublished doctoral dissertation, University of Wisconsin–Madison, 1981.

Carpenter, T. P., Corbitt, M. K., Kepner, H., Lindquist, M. M., & Reys, R. E. Results and implications of the second NAEP mathematics assessment: Elementary school. *Arithmetic Teacher,* 1980, *27,* 10–12, 44–47.

Carpenter, T. P., Hiebert, J., & Moser, J. M. First-grade children's initial solution processes for simple addition and subtraction problems. *Journal for Reserch in Mathematics Education,* 1981, *12,* 27–39.

Carpenter, T. P., & Moser, J. M. *An investigation of the learning of addition and subtraction.* (Theoretical Paper No. 79). Madison: Wisconsin Research and Development Center for Individualized Schooling, November 1979.

Carpenter, T. P., Moser, J. M., & Hiebert, J. *The effect of instruction on first-grade children's solutions of basic addition and subtraction problems.* (Working Paper No. 304). Madison: Wisconsin Research and Development Center for Individualized Schooling, 1981.

Gibb, E. G. Children's thinking in the process of subtraction. *Journal of Experimental Education,* 1956, *25,* 71–80.

Greeno, J. G. *Some examples of cognitive task analysis with instructional implications.* Paper presented at the ONR/NPRDC Conference, San Diego, Calif., March 1978.

Groen, G. J., & Parkman, J. M. A chronometric analysis of simple addition. *Psychological Review,* 1972, *79,* 329–343.

Groen, G., & Resnick, L. B. Can preschool children invent addition algorithms? *Journal of Educational Psychology,* 1977, *69,* 645–652.

Grouws, D. A. Differential performance of third-grade children in solving open sentences of four types. (Doctoral dissertation, University of Wisconsin, 1971.) *Dissertation Abstracts International,* 1972, *32,* 3860A.

Hebbeler, K. Young children's addition. *The Journal of Children's Mathematical Behavior,* 1977, *1*(4), 108–121.

Hiebert, J. *Young children's solution processes for verbal addition and subtraction problems: The effect of the unknown set.* Paper presented at the annual meeting of the National Council of Teachers of Mathematics, St. Louis, 1981.

Jerman, M. Problem length as a structural variable in verbal arithmetic problems. *Educational Studies in Mathematics,* 1973, *5,* 109–123.

Lankford, F. G. What can a teacher learn about a pupil's thinking through oral interviews? *Arithmetic Teacher,* 1974, *21,* 26–32.

Lindvall, C. M., & Ibarra, C. G. Incorrect procedures used by primary grade pupils in solving open addition and subtraction sentences. *Journal for Research in Mathematics Education,* 1980, 11(1), 50–62.

Nesher, P. S., & Katriel, T. *Two cognitive modes in arithmetic work problem solving.* Paper presented at the second annual meeting of the International Group for the Psychology of Mathematics Education, Osnabruck, West Germany, September 1978.

Riley, M. S. *The development of children's ability to solve arithmetic word problems.* Paper presented at the annual meeting of the American Educational Research Association, San Francisco, 1979.

Rosenthal, D. J. A., & Resnick, L. B. Children's solution processes in arithmetic word problems. *Journal of Educational Psychology,* 1974, *66,* 812–825.

Steffe, L. P., Spikes, W. C., & Hirstein, J. J. *Summary of quantitative comparisons and class inclusion as readiness variables for learning first grade arithmetical content.* Athens, Ga.: The Georgia Center for the Study of Learning and Teaching Mathematics, University of Georgia, 1976.

Suppes, P., & Groen, G. Some counting models for first grade performance data on simple facts. In J. M. Scandura (Ed.), *Research in Mathematics Education,* Washington, D.C.: National Council of Teachers of Mathematics, 1967.

Suppes, P., Loftus, E. F., & Jerman, M. Problem solving on a computer-based teletype. *Educational Studies in Mathematics,* 1969, *2,* 1–15.

Woods, S. S., Resnick, L. B., & Groen, G. J. An experimental test of five process models for subtraction. *Journal of Educational Psychology,* 1975, *1,* 17–21.

3 Levels of Description in the Analysis of Addition and Subtraction Word Problems

Pearla Nesher
Haifa University and Goldie Rotman Center for Cognitive Science at the Hebrew University of Jerusalem

When one speaks of first steps in acquisition of addition and subtraction, does one have in mind the concrete operations with objects (i.e., collecting, adding, removing real objects, and keeping track of their quantity), or does one think of mathematical operations with numbers? The answer to this question is not always clear. There is no dispute concerning the fact that a child entering formal schooling (at the age of 5 or 6) has already gained a considerable amount of experience with real objects, collecting them, distributing them, removing them, etc. The dispute starts when one is evaluating the role of such experience in the child's future mathematical learning. The dispute reflects a difference in approach from an epistemological point of view.

In our world, the objects we use serve many purposes. When there is a change in their role, we also change the way we conceive them. For example, an apple is usually used for eating; however, sometimes apples are counted. Does this mean that counting and eating are the same operation? The existence of objects, by itself, does not lead to a unique way of looking at them, and the idea of abstracting some concepts from objects is not an obvious one. Although the way we observe objects is by means of abstract concepts (predicates in language), it is not clear at all that these concepts are abstracted from the objects. It might be considered that we observe and perceive objects through our already available conceptual frameworks.

If such a platonic view is taken, one cannot expect that a child, through manipulation of concrete objects by themselves, will arrive at mathematical notions like addition or subtraction. Some consider the activity of putting together to exemplify the addition operation. However, putting together does not always lead to addition. Let's remind ourselves of Popper's (1963) example of putting $2 + 2$ rabbits in a basket, where later on we find 7 or 8 in it. If we put $2 + 2$ drops into a dry flask, we can never get 4 out of it. Does this mean that $2 + 2$ is not 4? Not at all. It emphasizes that we are selective in choosing the facts we use to demonstrate the application of the arithmetic sentence $2 + 2 = 4$. As Popper (1963) later explains:

> . . . the equation $2 + 2 = 4$ only applies to objects to which nothing happens . . . to the extent to which our apples do not rot, or rot only very slowly, or to which our rabbits or crocodiles do not happen to breed; to the extent, in other words, to which physical conditions resemble the pure logical or arithmetical operation of addition, to the same extent, of course, does arithmetic remain applicable [p. 212].

Thus, we have a logical or arithmetic operation (or relation) on the one hand and facts or operations with objects in the real world on the other hand, and we hold some correspondence between selected aspects of each. Learning the rules of such correspondence is learning the application of addition and subtraction. In the framework of school, we usually refer to it as learning to solve word problems.

The real dispute is, what should be the starting point for teaching addition and subtraction? Should we start with numbers and their symbolic operation or with applications of addition and subtraction? One should note that the dispute is not one of concrete versus abstract, as both approaches employ concrete materials for exemplification, but rather an epistemological dispute, which dictates different starting points in the acquisition of mathematics in general, and addition and subtraction in particular. Different answers not only lead to different approaches concerning the beginning steps, but also illustrate that this is a complicated question that has a plurality of aspects, each of which should be studied in order to comprehend the entire phenomenon.

The research reported here has been conducted within the framework of the first approach, namely, first teaching the operations with numbers and then proceeding with the application of these operations to situations in the real world. The chapter summarizes a series of studies on solving word problems of addition and subtraction. The research was aimed at finding regularities in children's performance in solving word problems and interpreting these regularities with the aid of structured and controlled variables. Based on previous linguistic analysis, the controlled variables were mainly logical, semantic, and syntactic. We hope that studying these variables can serve as a first step toward any theory concerning the child's competence in mathematical thinking in general, and problem solving in particular.

Concentrating on addition and subtraction word problems seemed justified on two grounds: (1) it is the first time that the child faces applications of arithmetic at school and avoiding failure at this stage is essential to the further learning of arithmetic; and (2) solving word problems, in general, consists of transition from a verbal formulation to a formal formulation and to operating in the formal domain.

This chapter presents an overview of several studies that took place in the last 5 years in Israel; only the main findings are presented. References are made to more detailed papers.

REGULARITIES FOUND IN SOLVING WORD PROBLEMS OF ADDITION AND SUBTRACTION

The empirical data in which we have found the stable regularities are taken from a computer-based program in arithmetic. It includes a strand for word problems with a sample of about 500 types of word problems. The formulation of each type is fixed and the computer at random changes the specific numbers to be calculated. We were not interested in numerical computation so the range of numbers was limited. The students' main task was to decide which operation to perform. During the year 1979, we collected data from four elementary schools in four

TABLE 3.1
Proportion of Correct Answers for Four Populations on Five Word Problems

	Proportion of correct answers in each population			
Problem	*A*	*B*	*C*	*D*
Five children were in the playground. Two more children entered the playground. How many children are now in the playground?	.88 (1254)	.88 (216)	.90 (224)	.89 (164)
Ruth ate 3 apples and Rachel ate 2 apples. How many apples did Ruth and Rachel eat in all?	.77 (1960)	.73 (580)	.77 (439)	.75 (136)
Joseph and Roland had 7 marbles altogether. Two of them were Joseph's. How many of them were Roland's?	.52 (1826)	.46 (470)	.49 (293)	.41 (119)
How many crayons does Dan have now, if he received 7 crayons and lost 3 of them?	.72 (1206)	.71 (305)	.74 (238)	.79 (134)
Mother gave John 8 dollars. How many dollars are left if he has spent 3?	.94 (967)	.89 (222)	.89 (256)	.85 (287)

Note: Number of answers is given in parentheses.

different cities in Israel. The average number of answers for each of 70 word problems (35 addition and 35 subtraction) in our experiment was about 1500 answers. The intercorrelation among all four populations concerning the proportion of correct answers on each word problem ranged from .93 to .99. In other words, when a question was easy, it was easy in each of the four independent populations and the same was true for an average or for a difficult word problem. A sample of our results is given in Table 3.1.

Such empirical regularities found in a large population suggest that there are structural variables inherent in the text of the word problems that make them easy or difficult to understand. We are now looking for a theoretical model that will account for that observed phenomenon. In the following sections, the theoretically derived variables we are now studying are described.

The three main components in our analysis, each with its operational variables, are:

The logical structure, which includes the type of the arithmetic operation and superfluous information.

The semantic component, which consists of two components: the contextual constituent, that includes dynamic, static, and comparison texts, and the lexical constituent that includes verbal cues and fit.

The syntactic component which includes some surface structure variables such as length, number of sentences, position of the question, and order of the strings.

THE LOGICAL STRUCTURE

Word problems have a specific underlying logical structure. In Nesher and Katriel (1977), we formulated the logical conditions that addition and subtraction word problems must fulfill. Among them were:

Every addition and subtraction word problem consists minimally of *three underlying strings,* with special semantic dependencies among them. Two strings comprise the information component, and the third one is the question component. In addition problems the predicates in the information component should indicate the fact that the two sets of objects (or events) mentioned belong to disjoint sets, and the predicate of the question component refers to the union of these two subsets. Subtraction word problems must fulfill the same relations among the sets; however, the information component should refer to the union of the two sets and one of the subsets, whereas the question component refers to the remaining disjoint set. The burden of the semantic dependencies among the three strings rests on the special requirements on the interrelations between the predicated quantified arguments. We refer to it as the condition on agreement between numerically quantified arguments.

To illustrate what we mean by "the condition on the agreement between quantified arguments," consider the following problem:

> Five boys bought tickets to the show and the girls bought four tickets. How many. . . .

The quantified arguments are "five boys" and "four tickets." What will be the union of such two subsets or what can be the question component in a text such as this problem?

Generally speaking, if the strings comprising the text contain two- or three-place predicates, numerically quantified arguments can occur in either the logical subject or the logical predicate. The condition on the agreement between numerically quantified arguments sets the constraint by which, in all three strings of a problem text, the numerical quantifiers should bind either the logical subjects or part of the logical predicates, respectively. This condition is characteristic of addition (or subtraction) word problems. Our example violates this condition and therefore serves as an example of an ill-defined text. It should be noted that the logical conditions specified in this section do not hold for all types of arithmetical word problems. In order for a problem text to be interpreted as calling for another arithmetic operation, different conditions have to be specified.

These logical conditions should appear even in the most minimal text word problem, otherwise the problem is insoluble. It may well be that a text contains more than the minimal information mentioned in the foregoing. In this case we refer to the additional information that does not fit this minimal structure (of two disjoint sets and their union) as superfluous information. The label *Superfluous Information* or *Extraneous Information* emphasizes the existence of a well-formulated logical structure.

Empirical Findings

It is already well established in the literature that a specific arithmetic operation is a structural variable in solving word problems (Jerman & Rees, 1971–72, Suppes, Loftus, & Jerman, 1969). For example, we found that average addition word problems are significantly easier than are subtraction word problems (with a mean of 73.9 for addition word problems and 64.6 for subtraction word problems).

The variable Superfluous Information has already been studied empirically by Cruickshank (1948); Goodstein, Cawley, Gordon, and Helfgott (1971); Goodstein, Bessant, Thibodeau, Vitello, and Vlakokos (1972), and Nesher (1976) and found to be significant. It is an indirect variable for learning about the role of the logical structure. When a word problem gives superfluous numerical information, the condition about the agreement between quantified arguments has to be clarified before the correct solution can be found. Let us consider the following example:

> Five boys bought three tickets, and the girls bought four tickets. How many tickets did the boys and girls buy?

When faced with such a problem, one should notice that there are three quantified arguments in the text: "five boys," "three tickets," and four tickets. The question is, on which numbers should one operate (assuming a binary operation), and what is the required operation? When processing this word problem text one might see two options: First, the disjoint sets are "boys" and "girls," and their union is boys and girls; or second, the disjoint sets are three tickets and four tickets (the fact that they are two different, disjoint sets is expressed by mentioning tickets bought by the boys and tickets bought by the girls), and the union set is tickets bought by the boys and the girls.

By applying the condition of agreement between the quantified arguments, the first option is excluded, because the subset "girls" is not quantified, and the number assigned to the union cannot be computed. Therefore, of all kinds of possible subsets and their union, only the second option can be further pursued, and the quantification of the subset "boys" should be understood as a superfluous numerical information that interferes with the minimal structure of an addition word problem.

THE SEMANTIC COMPONENT

As a result of previous analysis (Nesher & Katriel, 1978; Nesher & Teubal, 1974), two different constituents are involved in a semantic analysis.

The Contextual Constituent

By the contextual constitutent we mean the nature of the text as a whole and the semantic dependencies among all its underlying strings. The agreement between the quantified arguments among the strings in the word problem text is encoded by a variety of linguistic mechanisms, detailed later.

The Lexical Constituent

By the lexical constituent we mean the effect of isolated lexical items appearing in the text. It has been noted (Dahmus, 1970; Jerman & Mirman, 1974; Linville, 1976; Nesher & Teubal, 1974; Searle, Lorton, & Suppes, 1974; Wheat & Kulm,

1976) that certain words when appearing in a word problem facilitate or hinder the child's performance in solving the problem. Some of the words were even coined in the literature as cue verbs versus distractors. What is characteristic of these words is that in many cases they are optional in the text and may be introduced, replaced, or removed in isolation (except for the case of word problems of the type Relational Terms) and therefore do not serve as part of the semantic dependencies characteristic of the contextual constituent.

The Underlying Contextual Constituent

In a semantic analysis of addition and subtraction word problems (Nesher & Katriel, 1977), it was shown that the logical relations expressed in the texts can be linguistically encoded in a variety of manners. Because we wanted to base the partition of our sample of word problems on semantic categories, the following seven dependence relationships were examined:

The italic words will emphasize the core of the semantic dependencies among the three strings of the text.

Arguments. Semantic dependence among the numerically quantified *arguments* occurring in the strings underlying the problem text. For example:

Three *boys* and two *girls* went to the beach. How many *children* went to the beach?

Adjective: Semantic dependence due to *adjectives* that qualify the quantified arguments. For example, *big* and *small* qualify the *arguments* in the problem:

There are four *big windows* and three *small windows* in the hall. How many *windows* are in the hall?

Agents. Semantic dependence due to the *agents* referred to in the text. For example, Ruth and Dina are the agents in:

Ruth had three apples and *Dina* had two apples. How many apples did *Ruth and Dina* have altogether?

Location. Semantic dependence due to the *spatial relation* among objects. For example, bed, shelf, and room locate the objects in:

There are two books on the *bed* and eight books on the *shelf*. How many books are in the *room* altogether?

Time. Semantic dependence due to the *temporal relation* among events referred to in the text. For example, yesterday and today are time referents in:

Dan ate three candies *yesterday* and two candies *today*. How many candies did Dan eat on *both these days?*

Verbs. Semantic dependence expressed in terms of the *verbs* occurring in the text. For example, had, gave, and have are the verbs in:

Victor *had* five stamps and *gave* two of them to Joe. How many stamps does Victor *have* now?

Relational terms. Semantic dependence due to relational terms concerning two given quantified arguments. For example, *more—than* are relational terms in:

Bill has 8 marbles. Tom has 5 *more* marbles *than* Bill. How many marbles does Tom have?

One should note the difference between the categories Argument and Agent. The mechanism describing the dependency in Argument is by the use of superordinate nouns. Mentioning Agents in addition to the quantified Arguments introduces numerous verbs of possession (Bendix, 1966; Dixon, 1973; Fillmore, 1968; Gentner, 1975; Schank, 1973). Texts with Relational Terms are different and pose a special difficulty (Greeno, 1978; Paige & Simon, 1966).

In addition to these dependence relationships three main contextual constituent categories for texts are important (Greeno, 1978; Moser, 1979; Vergnaud & Durand, 1976):

Dynamic Texts. These are word problems in which the addition and subtraction logical relations are embedded in a temporal sequence of events. Thus, one can distinguish occurrences at different times ($T_1 < T_2 < T_3$). Following Lyons (1977) and Lipsky (1974), one should note that introducing change at T_2 *is the core* of a dynamic text. Something must have occurred in order to make the text a dynamic one. However, just mentioning different time points in a text by itself does not make the text a dynamic one. For example the text: "Five children went to the movies this week. Three of them went on Sunday and the rest went on Monday. How many children went to the movies on Monday?" is *not* a dynamic text, even though T_1 (Sunday) and T_2 (Monday) are mentioned. The time points mentioned serve to indicate the disjointedness of the two sets of children, but do not describe any change in the event described in T_1. Thus, for a text to be a dynamic one, some manipulation of the set of objects referred to in T_1 is necessary. Also, dynamic texts are mainly characterized by the action described by means of verbs, rather than by time points. We have had difficulty in formulating a dynamic *addition* word problem. Maybe it is due to the addition's logical structure, by which inherently one does not operate on the first-mentioned set.

Static Texts. These word problems, in which the addition or subtraction logical relations are encoded by means of Arguments, Adjectives, Agents, Location, and Time (points), will be considered as static word problems.

Comparison Texts. In these word problems two sets of objects are compared on the same attribute. The linguistic device for such a comparison involves the use of words such as more or less, and at least one relational quantity (i.e., not given in its absolute value).

Empirical Glimpses

For the sample of 35 addition and 35 subtraction word problems mentioned earlier, the mean percentage correct for each of the forms of dependence relationship for both addition and subtraction word problems are shown in Table 3.2.

The analysis of *subtraction texts* yielded three distinct contextual categories:

Static, which includes the categories Arguments, Adjectives, Agents, Location, and Time Points.
Dynamic, which includes the category Verbs.
Comparison, which includes the category of Relational Terms.

The analysis of the addition texts, however, ended with only two distinct categories: the Comparison text versus all the rest. Thus, word problems of addition containing Verbs denoting the disjointedness of the sets were not significantly different from the Static texts. For example, "Two boys ran . . . and three boys walked . . . ," although belonging to the category Verbs in our dependencies analysis, does not constitute a dynamic text. In this example, there is no change effect and no time order ($T_1 < T_2 < T_3$). Therefore, belonging to this specific category Verbs is not a sufficient condition for being a dynamic text. Table 3.3 presents the means for the contextual variables for the addition and subtraction word problems. Clearly, addition and subtraction word problems, despite the similarity in their logical structure and the fact that both employ the same linguistic mechanism to encode their logical structure, are different in nature. They differ not only in their means but also in the role the various semantic devices play in the expression of the logical structure (Steffe & Johnson, 1971). For addition word problems, the level of difficulty is not related to the contextual constituent (except for Comparison). However, for subtraction word problems, the semantic contextual constituent has a most influential role.

TABLE 3.2
Mean Percentage Correct for the Dependence-Relationship Categories for Addition and Subtraction Texts

Semantic Dependence	*Addition Word Problems*	*Subtraction Word Problems*
Arguments	80.03	52.56
Adjectives[a]	72.73	59.47
Agents	80.39	50.48
Time	77.90	38.00
Verbs	77.89	72.11
Comparison	57.32	60.14

[a]The category Location was combined with Adjective, as both were small samples and most similar in their means.

TABLE 3.3
Differences in Success with Addition and Subtraction Problems by Context

Contextual Variables	*Mean Success of Addition*	*Mean Success of Subtraction*
Static	77.55	51.71
Dynamic		74.83
Comparison	57.32	65.03
F-prob. in one-way ANOVA	0.00	0.00

Subtraction dynamic texts are much easier than static ones, and the comparison word problems are in between.

The Lexical Constituent

In previous studies the variable Verbal cues yielded different results when studied in two different settings. In one study on addition and subtraction comparison problems, the Verbal cue variable was found to be significant (Nesher & Teubal, 1974). Later, however, when this variable was incorporated into a study in which two additional structural variables were presented (Superfluous Information and Number of Binary Operations), Verbal cue was found to be not significant.

In our present study, we decided not to judge which word will serve as a verbal cue and which word will serve as a distractor a priori. Rather, we labeled the lexical relevant items in the following manner. First, words that describe the union sets (in all, altogether, etc.) and increasing (more) in any addition and subtraction word problems were labeled as +CUE. Second, words that describe subsets (of them, the rest, were left, remained, etc.) and decreasing (less) were labeled as −CUE.

A third label, FIT, was also used. Verbal cues were also numbered: 1 when the lexical item was in agreement with the logical set it referred to; 0 when the lexical item was irrelevant to the logical structure; and −1 when the lexical item was in contradiction to the logical set it referred to. Then combinations of the values +CUE, −CUE, and FIT in addition and subtraction word problems were examined separately for each contextual variable. The regression coefficients from a multiple regression analysis, and the total variance explained by the variables +CUE and −CUE for the contextual variables are as shown in Table 3.4.

The major findings were that the contextual semantic constituent is stronger than the lexical one, though both were significant. The lexical constituent has almost no role in Dynamic subtraction word problems, whereas it has a very strong effect in Comparison texts, and a moderate role in Static texts. The same

TABLE 3.4
Regression Coefficients and Total Variance for +CUE and −CUE

	Lexical Variable	*Regression Coefficients*			
		All	*Dynamic*	*Static*	*Comparison*
Subtraction	+CUE	−8.84	−4.60	−9.52	−6.60
	−CUE	2.82	0.51	2.07	16.56
		R Square			
		0.59	0.04	0.23	0.77
		Regression Coefficients			
		All	*Dynamic and Static*		*Comparison*
Addition	+CUE	[a]	6.57		
	−CUE	−8.83′	[a]		28.83
		R Square			
		All	*Dynamic and Static*		*Comparison*
		0.34	0.13		0.96

[a]Too small to enter the regression equation

is true with addition problems. The lexical constituent has a crucial role in Comparison texts, but a small effect in Dynamic and Static texts. The dominant role of the Verbal cue in Comparison texts may be due to the fact that the verbal cue in such texts is not optional, but rather an integral part of the logical structure. The regression coefficients assigned to +CUE and −CUE in each case demonstrate that their contribution in each context is not symmetric, and they should be weighted accordingly. Also, our first explorations concerning the label FIT show that it is a moderating variable and therefore, attributes a more accurate weight to the lexical items.

THE SYNTACTIC COMPONENT

Surface Structure Variable

Several studies dealing with linguistic aspects of word problems in arithmetic devoted their efforts to looking into the influence of surface structure syntactic variables (Jerman, 1973; Jerman & Rees, 1971–1972; Linville, 1976). We also included a few syntactical variables in our research and all of them were found to be not significant. The variables analyzed were:

1. *Number of sentences*. Whether the word problem consists of one, two, or three sentences.
2. *Location of the question*. Whether the question is in the beginning, middle, or end of the text.
3. *Number of words*.

Another surface variable examined in an earlier study was the order of the strings and whether or not this order reflects the order of events in dynamic texts. This too was not significant (Nesher & Katriel, 1978). These results concerning surface structure syntactic variables are not surprising and are in accord with general linguistic analysis.

FINAL REMARKS

We regard problem solving in arithmetic as a complex set of cognitive tasks. Each task must be fully understood before the regularities found in children's performance can be accounted for. We have tried in a series of studies to employ three different levels of description, each of which consists of several variables. The main levels of analysis we explored were the logical structure of the text, the semantic interpretation of the text, and the syntactic component of the text. We have found that when the logical structure of the word problem is controlled, the contextual constitutent of the semantic variable is most influential, and that the lexical component of isolated words can be fully appreciated only in interaction with the more general contextual semantic variables. We have also progressed in our ability to weight the verbal cues. Finally, surface structure variables have no significant effect when the task of solving addition and subtraction word problems is examined.

However, we have not exhausted the variables affecting the performance of solving word problems. More are now under study. One is the role of dynamic texts in the logical structure of addition. Another variable is related to the problem we call "the temporal sequentiality of the events, and the flow of information in the verbal text." This also deals with the canonical versus noncanonical numerical sentences that are chosen as solutions for a given verbal problem.

In summary, it is our conviction that the complex tasks of arithmetic can only be fully understood in terms of their logical structure and their linguistic structure in relationship to psychological models for internal representation of symbolic information. As we gain better understanding of these factors, reasonable answers to the problem of understanding how children solve arithmetic word problems will be obtained.

REFERENCES

Bendix, H. *Componential analysis of general vocabulary*. Bloomington: Indiana University, 1966.

Cruickshank, W. Arithmetic ability of mentally retarded children: Ability to differentiate extraneous material from needed arithmetic facts. *Journal of Educational Research,* 1948, *42,* 161–170.

Dahmus, M. How to teach verbal problems. *School Science and Mathematics,* 1970, *70,* 121–138.

Dixon, R. The semantics of giving. *The formal analysis of natural languages*. The Hague: Mouton, 1973.

Fillmore, C. Lexical entries for verbs. *Foundations of Language,* 1968, *4,* 373–393.

Gentner, D. Psychological reality of semantic components: Verbs of possession. In D. Norman and D. Rumelhart (Eds.), *Exploration in cognition*. San Francisco: Freeman & Co., 1975.

Goodstein, H. A., Bessant, H., Thibodeau, G., Vitello, S., & Vlakokos, I. The effect of three variables on the verbal problem solving of educable mentally handicapped children. *American Journal of Mental Deficiency,* 1972, *76*(6), 703–709.

Goodstein, H. A., Cawley, J. F., Gordon, S., & Helfgott, J. Verbal problem solving among educable mentally retarded children. *American Journal of Mental Deficiency,* 1971, *76*(2), 238–241.

Greeno, J. G. *Some examples of cognitive task analysis with instructional implications*. Paper presented at ONR/NPRDC Conference, San Diego, March 6–9, 1978.

Jerman, M. Problem length as a structural variable in verbal arithmetic problems. *Educational Studies in Mathematics,* 1973, *5,* 109–123.

Jerman, M., & Mirman, S. Linguistic and computational variables in problem solving in elementary mathematics. *Educational Studies in Mathematics,* 1974, *5,* 317–362.

Jerman, M., & Rees, R. Predicting the relative difficulty of verbal arithmetic problems. *Educational Studies in Mathematics,* 1971–1972, *4,* 306–323.

Linville, W. Syntax, vocabulary, and the verbal arithmetic problems. *School Science and Mathematics,* 1976, *76,* 152–158.

Lipsky, J. Towards a topology of semantic dependence. *Semiotica,* 1974, *12,* 2.

Lyons, J. *Semantics*. London: Cambridge Press, 1977.

Moser, J. M. *Young children's representation of addition and subtraction problems*. Conceptual Paper No. 4, Madison: Wisconsin Research and Development Center for Individualized Schooling, 1979.

Nesher, P. Three determinants of difficulty in verbal arithmetic problems. *Educational Studies in Mathematics,* 1976, *7,* 369–388.

Nesher, P., & Katriel, T. A semantic analysis of addition and subtraction word problems in arithmetic. *Educational Studies in Mathematics,* 1977, *8,* 251–269.

Nesher, P., & Katriel, T. *Two cognitive modes in arithmetic word problem solving*. Paper presented at the second annual meeting of the International Group for the Psychology of Mathematics Instruction, Osnabruck, West Germany, 1978.

Nesher, P., & Teubal, E. Verbal cues as an interfering factor in verbal problem solving. *Educational Studies in Mathematics,* 1974, *6,* 41–51.

Paige, J., & Simon, H. Cognitive processes in solving algebra word problems. In B. Kleinmuntz (Ed.), *Problem solving*. New York: John Wiley & Sons, 1966.

Popper, K. *Conjectures and refutations: The growth of scientific knowledge*. New York: Harper & Row, 1963.

Schank, R. *The fourteen primitive actions and their inferences*. Stanford Artificial Intelligence Laboratory, Memo AIM-183, 1973.

Searle, B. W., Lorton, P. Jr., & Suppes, P. Structural variables affecting CAI performance on arithmetic word problems of disadvantaged and deaf students. *Educational Studies in Mathematics,* 1974, *5,* 371–384.

Steffe, L., & Johnson, D. Problem-solving performances of first-grade children. *Journal for Research in Mathematics Education,* 1971, *2,* 50–64.

Suppes, P., Loftus, E., & Jerman, M. Problem solving on a computer-based teletype. *Educational Studies in Mathematics,* 1969, *2,* 1–15.

Vergnaud, G., & Durand, C. Structures additives et complexité psychogénétique. *La Revue Francaise de Pedagogie,* 1976.

Wheat, M., & Kulm, G. *The effects of vocabulary and translation training on solving verbal problems.* Paper read at a NCTM Research Session: Atlanta, Georgia, 1976.

4 A Classification of Cognitive Tasks and Operations of Thought Involved in Addition and Subtraction Problems

Gérard Vergnaud
Centre d'Etude des Processus Cognitifs et du Langage

The purpose of this chapter is to present and explain a classification of additive relationships helpful in interpreting the procedures students use in solving addition and subtraction problems and in understanding more about the difficulties they meet. It also provides a framework for understanding the meaning of different symbolic representations of addition and subtraction, and a coherent basis for designing experiments on these mathematical processes. The theoretical reasons for the distinctions in the classification scheme are both psychological and mathematical. Let me illustrate with three examples of problems:

Problem A. There are 4 boys and 7 girls around the table. How many children are there altogether?

Problem B. John just spent 4 francs. He now has 7 francs in his pocket. How much did he have before?

Problem C. Robert played two games of marbles. On the first game, he lost 4 marbles. He played the second game. Altogether, he now has won 7 marbles. What happened in the second game?

Although a simple addition, 4 + 7, is needed in all three cases, Problem B is solved 1 or 2 years later than A, and C is failed by 75% of 11-year-old students. There must be some logical or mathematical difficulties in the last two problems that do not exist in the first.

To study this sort of hierarchy, I believe that a psychogenetic approach would be valuable. Until now psychogenetic theory has not been applied to specific content areas. Its use has been restricted to general problems of development, whereas study of specific content has been the object of learning theory. I consider this division to be somewhat misleading. It would be fruitful for education to consider a synthesis of psychogenesis and learning. One way to construct such a synthesis is to consider that knowledge is organized in "conceptual fields," the mastery of which develops over a long period of time through *experience, maturation,* and *learning*. By conceptual field, I mean an informal and heterogeneous set of problems, situations, concepts, relationships, structures, contents, and operations of thought, connected to one another and likely to be interwoven during the process of acquisition. For example, the concepts of multiplication, division, fraction, ratio, proportion, linear function, rational number, similarity, vector space, and dimensional analysis all belong to one single large conceptual field, the field of "multiplicative structures."

Similarly, I believe that the concepts of measure (of discrete sets and of other magnitudes), addition, subtraction, time transformation, comparison relationship, displacement and abscissa on an axis, and natural and directed number are also elements of one single conceptual field, the field of "additive structures." The progressive understanding of that field develops over a long period of time, from the age of 3 or 4 until at least age 15 or 16. The psychogenetic study of the acquisition of that field requires the analysis of the different relationships involved, and the hierarchical study of the different classes of problems that may be offered to students. It requires also the study of the different procedures and the different symbolic representations that students may use.

Children usually build up a conceptual field through experience in daily life and school. It is fruitful to plan and perform didactic experiments in order to understand more about how the field is constructed. For such experiments, it is essential to know which structures and classes of problems are the most easily understood by young students, which are the next, and so on. We must also know which procedures are the most naturally used by children or the most easily assimilated when taught. The same is true for symbolic representations: diagrams of different kinds, equations of different sorts. Such studies should enlighten our view of the slow process of acquisition and give us a better understanding of children's behavior.

To interpret the behavior of children faced with elementary arithmetic problems, I find it essential to distinguish two sorts of calculus: "relational calculus" and "numerical calculus." By "numerical calculus" I mean ordinary operations of addition, subtraction, multiplication, and division. By "relational calculus" I mean the operations of thought that are necessary to handle the relationships involved in the situation. This calculus can usually be expressed in terms of theorems, when it is valid, or in terms of false inferences, when it is not. But these theorems, assumptions, and inferences are not necessarily expressed or explained by children; they can only be hypothesized by observing children's actions. Let us call them "theorems in action" or "inferences in action."

If one wants to take such inferences into account, one has to develop a mathematical frame of reference of elementary arithmetic problems that includes aspects of the problem situations that are not usually taken into account by mathematicians or by text authors. For example mathematicians are not in-

TABLE 4.1
Relational Calculus and Numerical Calculus

Illustrative Examples	*Diagram and Relational Calculus*	*Numerical Calculus*
F+ Fred had 3 sweets, he buys 14 sweets. How many sweets does he have now?	[3] —(+14)→ [ ] apply a direct positive transformation to the initial state	addition
F− Fred had 17 sweets, he eats 4 of them. How many sweets does he have now?	[17] —(−4)→ [ ] apply a direct negative transformation to the initial state	subtraction
T+ Tony had 5 marbles. He plays with Robert. He has now 12 marbles. What has happened during the game?	[5] —( )→ [12] find the difference between two states I < F	subtraction
T− Tony had 7 marbles. He plays with Robert. He now has 2 marbles. What has happened during the game?	[7] —( )→ [2] find the difference between two states I > F	subtraction
I+ Inge has just received 4 dollars from her mother. She now has 7 dollars. What did she have before?	[ ] —(+4)→ [7] find the inverse of a positive transformation and apply it to the final state.	subtraction
I− Inge has just spent 3 dollars to buy sweets. She has now 8 dollars. What did she have before?	[ ] —(−3)→ [8] find the inverse of a negative transformation and apply it to the final state.	addition

terested in the concepts of time and dimension. But children take time aspects and dimensional aspects into consideration. One can hardly understand what they are doing if one keeps these aspects out of the frame of reference. To illustrate, let us take Problem D and Problem E:

> Problem D: 4 boys, 7 girls; how many children are there altogether?

This exemplifies the classical measure–measure–measure relationship: One measure is the composition of two elementary measures. There are only two classes of such problems: first, find the sum knowing the elementary measures, and second, find one elementary measure knowing the sum and the other elementary measure. These two classes of problems are in one-to-one correspondence with the numerical operations of addition and subtraction.

> Problem E: John has just spent 4 francs, he now has 7 francs; how much did he have before?

This exemplifies a different relationship, a measure–transformation–measure relationship. From these examples the six distinct classes of problems can be generated (Table 4.1). However, these six classes of problems are not in a simple correspondence with the numerical operations of addition and subtraction.

If problem solving is both the source and the basis of operational knowledge, it is essential to use a classification that encompasses all (or most) classes of problems and most aspects of problem solving, rather than a framework based only on the numerical operations (addition, subtraction) and on the concepts of number and equation, which fail to map many relevant features of the cognitive tasks involved. This does not mean that the concepts of number and equation are not essential. However, they are abstract entities that elementary school children find difficult to handle. The reason is that they are a condensation of too many different relationships and entities.

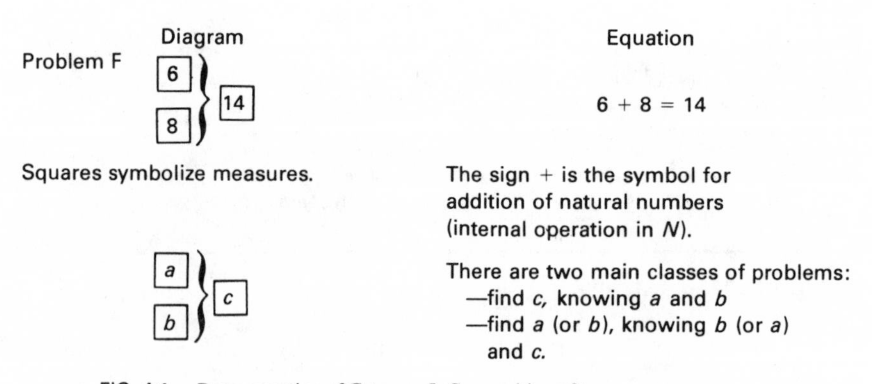

FIG. 4.1. Representation of Category I: Composition of two measures.

BASIC CATEGORIES OF RELATIONSHIPS

To represent the six main categories of relationships that I have found necessary to distinguish, let me use special symbols and notation. The six categories are explained in Fig. 4.1 through 4.6 and the notation is summarized in Fig. 4.7.
Category I: Composition of two measures. Problem F illustrates this category:

> Problem F: Peter has 6 marbles in his right-hand pocket and 8 marbles in his left-hand pocket. He has 14 marbles altogether.

The two classes of problems in this category are mentioned in Fig. 4.1.
Category II: A transformation links two measures (state-transformation-state [STS]). Problem G illustrates this category:

> Problem G: Peter had 17 marbles before playing. He has lost 4 marbles. He now has 13 marbles.

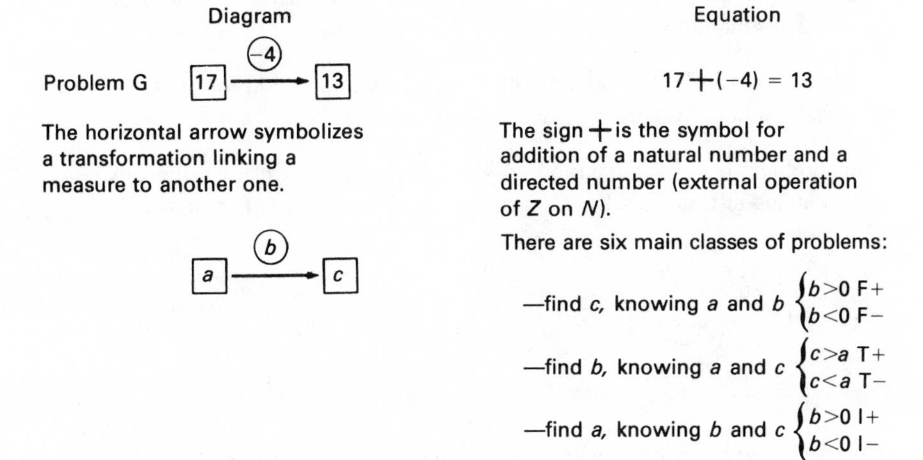

FIG. 4.2. Representation of Category II: State-Transformation-State (STS).

The six classes of problems (F+, F−, T+, T−, I+, I−) outlined in Table 4.1 and in Fig. 4.2 are in this category. Category II refers to the concept of a transformation happening in time.
Category III: A static relationship links two measures (state-relationship-state [SRS]). Problem H illustrates this category:

> Problem H: Peter has 8 marbles. He has 5 more than John. John has 3 marbles.

Figure 4.3 represents this category, which has also six classes of problems, analogous to those of Category II. I have found it necessary to distinguish this category from Category II to highlight the difference between dynamic transformations and static relationships.

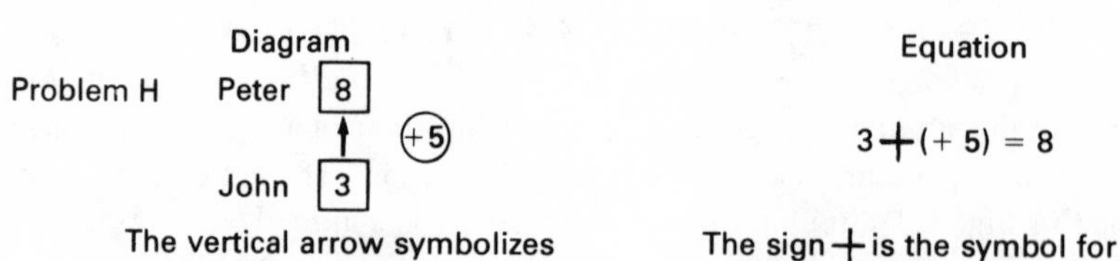

The vertical arrow symbolizes a static relationship (a comparison in this case) linking two measures.

The sign + is the symbol for addition of a natural number and a directed number as in category II.

There are six main classes of problems (analogous to those described in Fig. 4.2).

FIG. 4.3. Representation of Category III: State-Relationship-State (SRS).

Category IV: Composition of two transformations (transformation-transformation-transformation [TTT]). Problem I illustrates this category.

> Problem I: Peter won 6 marbles in the morning. He lost 9 marbles in the afternoon. Altogether he lost 3 marbles.

This problem is diagrammed in Fig. 4.4. There are many classes and subclasses of problems in Category IV. To generate these classes see Fig. 4.4.

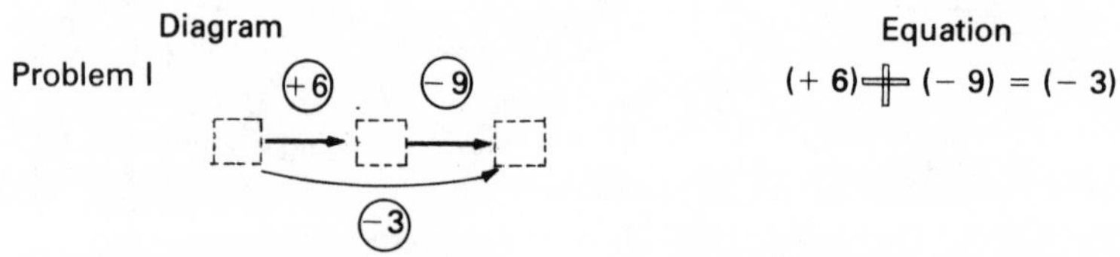

The lower arrow symbolizes the composition of the two upper ones. The states are not known.

The sign + is the symbol for addition of two directed numbers (internal operation in Z).

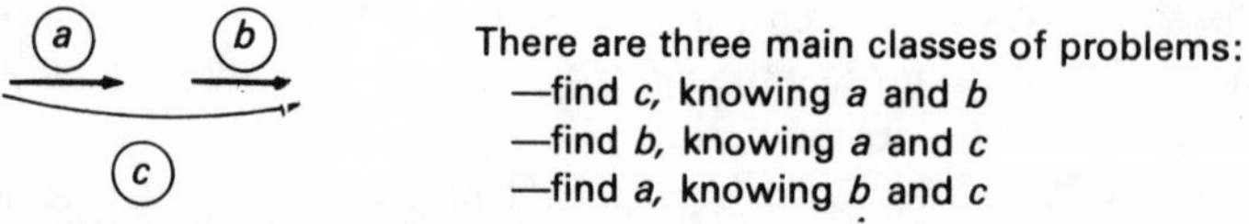

There are three main classes of problems:
- —find c, knowing a and b
- —find b, knowing a and c
- —find a, knowing b and c

Within each of these main classes, there are subclasses depending on the values of the data. For instance, if a and c are given, one gets the following cases:

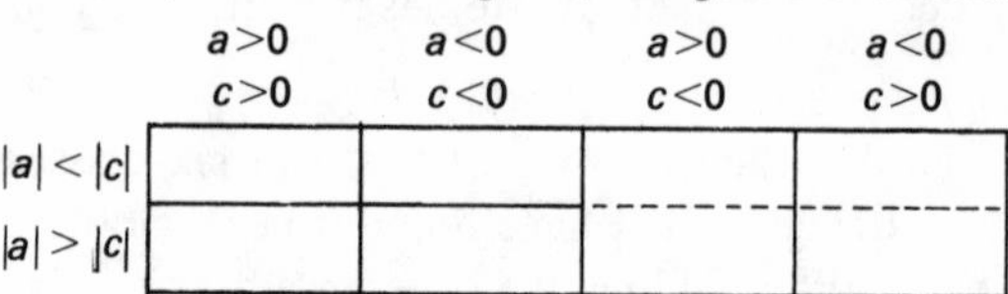

	$a>0$, $c>0$	$a<0$, $c<0$	$a>0$, $c<0$	$a<0$, $c>0$
$\|a\|<\|c\|$				
$\|a\|>\|c\|$				

FIG. 4.4. Representation of Category IV: Composition of two transformations (TTT).

Category V: A transformation links two static relationships (relationship-transformation-relationship [RTR]). Problem J illustrates this category.

> Problem J: Peter owed Henry 6 marbles. He gives him 4. He still owes Henry 2 marbles.

Different diagrams and equations are possible in this case (see Fig. 4.5).

Diagram — Equation

Problem J

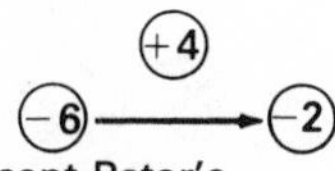

$(-6) \oplus (+4) = (-2)$

States represent Peter's account in Henry's book of credit.

The sign $\oplus$ is the symbol for addition of two directed numbers. It actually stands in this case for an external operation of transformations on states.

Another diagram could be:

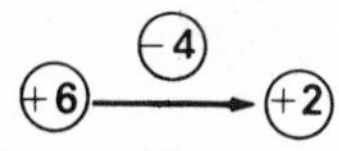

States would then represent Henry's account in Peter's book.

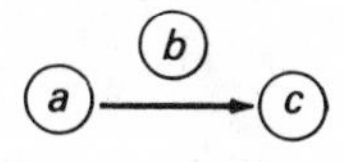

There are three main classes of problems as in Category IV. Each one can also be subdivided into different subclasses.

FIG. 4.5. Representation of Category V· Relationship-Transformation-Relationship (RTR).

Category VI: Composition of two static relationships (relationship-relationship-relationship [RRR]). Two problems will illustrate this category.

> Problem K: Peter owes 8 marbles to Henry, but Henry owes 6 to Peter. So Peter owes 2 marbles to Henry.
>
> Problem L: Robert has 7 marbles more than Susan. Susan has 3 marbles less than Connie. Robert has 4 marbles more than Connie.

This category is represented in Fig. 4.6. Because there is no fixed time order, there are several possible representations of the same situation.

The classification of problems into these six main categories is comprehensive. This classification relies on the distinction between three main concepts; measure, time transformation, and static relationship.

In this regard I would like to stress two points: First, time transformations and static relationships are not adequately represented by natural numbers, and problems that involve them are not adequately represented by equations in N. N is adequate for measures (of discrete sets) and for Category I relationships. It is not

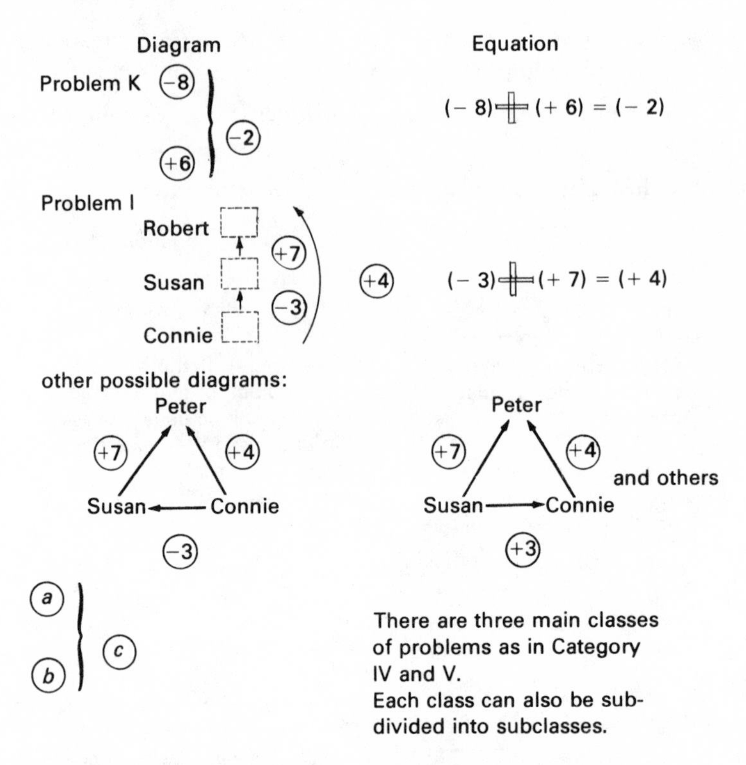

FIG. 4.6. Representation of Category VI: Relationship - Relationship - Relationship (RRR).

suitable for Category II, III, IV, V, or VI relationships because they involve elements that should be represented by directed numbers. However, students meet these categories long before they learn about directed numbers. Thus, there is a discrepancy between the structure of problems that children meet and the mathematical concepts that they are taught.

Second, the term "dynamic" is ambiguous. It may refer to the action of an actor or it may refer to a transformation that has nothing to do with any action. Initially, I was interested in the idea that because action plays a central role in building concepts, teaching children a dynamic model would be more productive than teaching a static one. My present view is more careful: One must distinguish carefully between the concept of *action,* the concept of *transformation,* and the concept of *operation*. Where action refers to an actor's doing, transformation refers to a change in the state of nature, and operation refers to the procedure used to solve a problem. Let us take three examples:

> John had 4 sweets. He buys 3 sweets. How many sweets does he have now?

If a student solves the problem by saying, "$4 + 3 = 7$," then "$+3$" stands for John's action, or the transformation of John's collection, or Peter's operation to solve the problem.

> John had 4 sweets. His mother gives him 3 sweets. How many sweets does he have now?

Now if the student says, "$4 + 3 = 7$," then "$+3$" may stand for the transformation of John's collection or Peter's operation to solve the problem.

> John had 4 sweets. He has just eaten 3 sweets. How many did he have before?

Now if the student says, "$4 + 3 = 7$," then "$+3$" stands for Peter's operation to solve the problem. It does not represent John's action or the transformation of John's collection.

Suppose the student writes "$4 + 3 = \square$," or "$4 \xrightarrow{+3} \square$." For the first two problems, has he or she written an equation or a diagram of the problem, or has he or she just written down his or her procedure to solve the problem? It is impossible to decide. For the third problem, if the student writes "$4 + 3 = \square$," or "$4 \xrightarrow{+3} \square$," he or she has clearly written down his or her procedure and not an equation or a diagram. To depict these, he or she should have written "$\square - 3 = 4$," or $\square \xrightarrow{-3} 4$." Thus, for certain classes of problems it is possible to decide whether a symbolic expression represents the actual situation, or the procedure used to solve the problem.

Hence, it should be clear that the term "dynamic" is ambiguous because it refers to different concepts that may be or may not be congruent. The hypothesis that dynamic situations are easier than static ones may be true for some situations, but not for all, as is shown in the next section.

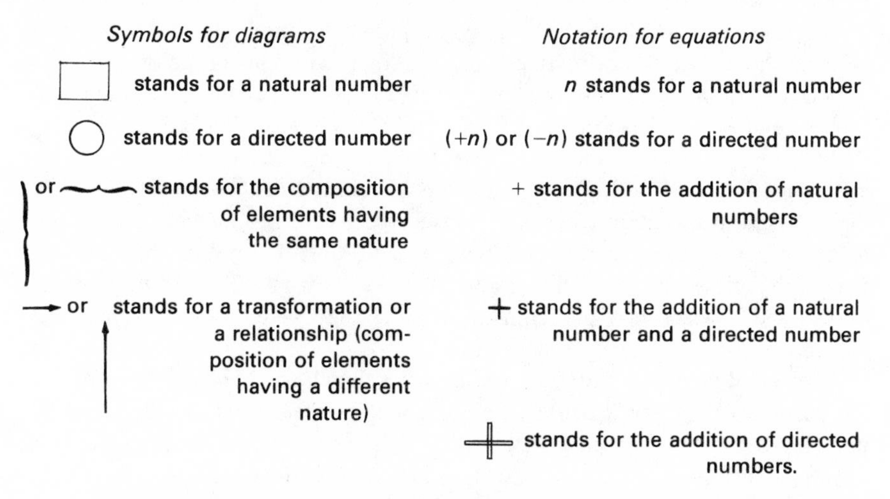

FIG. 4.7. Symbols and notation used to represent categories of addition and subtraction problems.

EXPERIMENTAL RESULTS

Experiment 1. In our first experiment (Vergnaud & Durand, 1976) we examined differences between Category II and Category IV problems and also differences between distinct classes of problems in Category IV. Because we stressed the relational aspects, small numbers were used. We used pairs of problems needing the same addition or subtraction equation but having different structures.

Table 4.2 shows pairs of problems and the number of correct responses for 28 children at each of five grades. There are some very clear differences. A 1-year decalage appears for the pair of problems Pierre and Paul, and at least a 3-year decalage for the pair Bertrand and Bruno. The decalage is smaller for the pair Claude and Christian. Although the same numerical calculus is needed in each pair of problems, the composition of transformations (Category IV) is more difficult than the application of transformations to states (Category II). However, the differences are not homogeneous for the different cases. For instance, the small difference between Christian and Claude may come from the fact that both cases can very easily be mapped into the composition-of-measures model, all transformations being positive. This is not the case with Pierre and Paul, nor with Bertrand and Bruno. The graphs show clearly the gap between adding two transformations and applying a transformation to a state. The gap is still bigger between finding the difference of two transformations having opposite signs and inversing a direct negative transformation.

Also, if one compares the problems in Category II, one finds some interesting differences. For instance, it is easier, in the case of a direct negative transformation, to calculate the final state (Pierre) than the initial state (Bertrand). There is roughly a 1-year decalage.

We also studied different Category IV problems, implying two successive transformations (games of marbles) and the composed transformation.

Starting from the TTT model shown in Fig. 4.4, we studied the class of problems illustrated by problem "Christian" (find *b,* knowing *a* and *c* or find T_2 knowing T_1 and T_3) and chose five problems out of the different possible cases shown in Table 4.3. The problem types again are identified in Table 4.3 by the names of the actors in each type. As we expected, Christian was found easier than Jacques, Jacques easier than Didier, and Christian easier than Didier. Cross-tabulations for those three situations are shown in Table 4.4. Further, Olivier and Vincent were particularly difficult. They were failed by 80% of the students of the last primary school grade.

The actual procedures used by students to solve these problbems are not just the canonical relational calculus. For instance, in Claude (Table 4.2) (find *T,* knowing the initial and final states when *T* is positive), we found not only a *difference* procedure (if *T* links two states, its value is the difference), but also a *complement* procedure (which consists of finding directly what should be added to the initial state to reach the final state).

TABLE 4.2
Comparison of Category II and Category IV Problems

Category II (STS) *State–Transformation–State*	*Category IV (TTT)* *Transformation–Transformation–Transformation*
Pierre has 6 marbles. He plays one game and loses 4 marbles. How many marbles does he have after the game?	*Paul* plays two games of marbles. At the first game, he wins 6 marbles. At the second game he loses 4 marbles. What has happened altogether?
Bertrand plays a game of marbles. He loses 7 marbles. After the game he has 3 marbles. How many marbles did he have before the game?	*Bruno* plays two games of marbles. He plays a first game, then a second game. At the second game he loses 7 marbles. After those two games he has won 3 marbles altogether. What has happened during the first game?
Claude has 5 marbles. He plays one game. After the game he has 9 marbles. What has happened during the game?	*Christian* plays two games of marbles. At the first game he wins 5 marbles. He plays a second game. After these two games he has won 9 marbles altogether. What has happened during the second game?

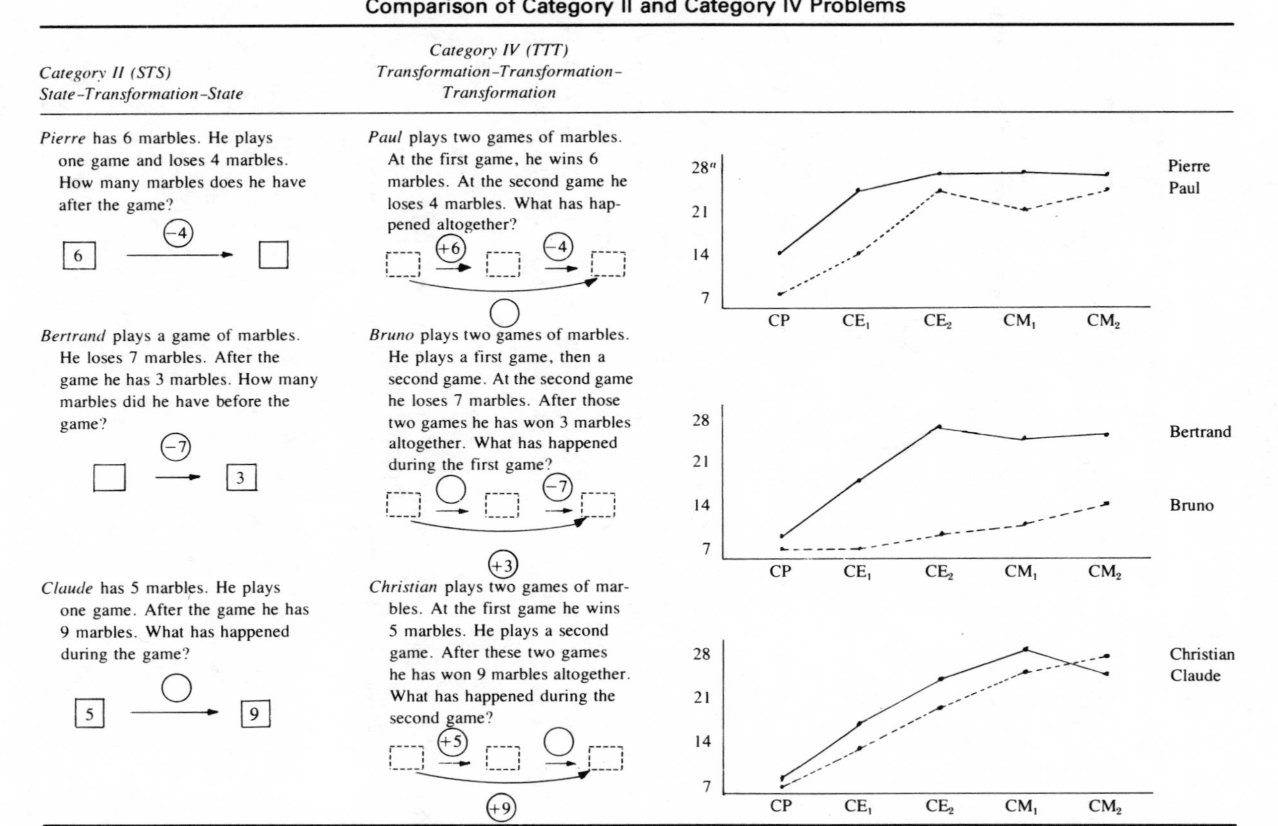

CP: first grade of French primary school (6-year-olds); CE_1, CE_2, CM_1, CM_2 are next grades.
[a]number correct

TABLE 4.3
TTT Five Problem Cases Studied in Vergnaud and Durand, 1976

	$T_1 > 0$, $T_3 > 0$	$T_1 < 0$, $T_3 < 0$	$T_1 > 0$, $T_3 < 0$	$T_1 < 0$, $T_3 > 0$
$\lvert T_3\rvert > \lvert T_1\rvert$	Christian	Jacques	Olivier	
$\lvert T_3\rvert < \lvert T_1\rvert$		Didier	Vincent	

Example: Christian plays two games of marbles. In the first game he wins 5 marbles ($T_1 > 0$). He plays a second game. After these two games he has won 9 marbles altogether ($T_3 > 0$ and $|T_3|>|T_1|$). What has happened during the second game?

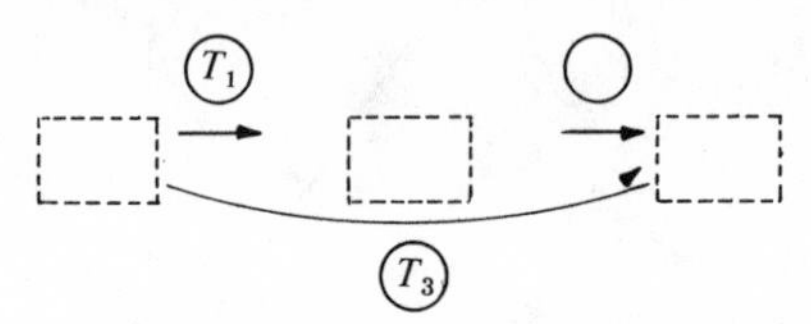

Similarly, in Bertrand (Table 4.2) (find the initial state, knowing T and the final state, when T is negative), we found not only the *inversion* procedure (take T^{-1} and apply it to the final state) but also the *hypothetical-initial-state* procedure (which is to make a hypothesis on the initial state, apply the direct transformation, find the outcome, compare with the actual final state, and either correct the hypothesis accordingly or make a new hypothesis). The *complement*

TABLE 4.4
Cross-Tabulation for Three Problem Pairs for the Category IV (TTT) Problem Cases

		Jacques C	Jacques I	n
Christian	C	72	19	91
	I	5	44	49
	n	77	63	140

		Didier C	Didier I	n
Jacques	C	49	26	75
	I	6	31	37
	n	55	57	112

		Didier C	Didier I	n
Christian	C	53	36	89
	I	2	21	23
	n	55	57	112

C = Correct
I = Incorret

procedure can be successfully used when T is positive. However for this case it leads to a failure.

In Category IV problems, *hypothesis* procedures are very often used because they require no inversion. Also, very often children interpret transformations as states. Sometimes, this interpretation enables the student to handle the problem, but often it is completely misleading.

Experiment 2. Conne (1979) repeated the first experiment and found essentially the same results. He did a detailed analysis of students' procedures and found that many procedures and explanations produced by children can be interpreted in terms of a functional model involving "what do I start from; what do I do next (add or subtract)?" This model is not suitable for all problems, but, in many cases it enables students to approach problems they would otherwise miss. Of course, using the functional model requires the identification of the starting point and the action.

For example, in T_3 problems (find T_3 knowing T_1 and T_2) the canonical solution requires the composition of two transformations. In fact, students very often consider that T_1 is the initial state and that T_2 operates on T_1. This interpretation can lead to success when T_1 and T_2 are positive, or when T_1 is positive and T_2 is negative and smaller in absolute value than T_1. But this model is not operational in other cases. In fact children's answers can be interpreted as inferences concerning the state of affairs before or after the games. For example: "wins 6 marbles" is interpreted as "he has 6 marbles," and "loses 7 marbles" is interpreted as "he had 7 marbles and he has no marbles left."

In problem Paul (Table 4.2) the answer is often given as the value of the final state: "Two marbles are left." In Vincent (Table 4.3), T_3 is often considered as operating on T_1: "Six marbles are left." This interpretation depends on the difficulty of the numerical values chosen. The same text may be interpreted in terms of transformations (correct) when the numerical values are easily handled (problems Christian and Jacques, Table 4.3), for instance, in which T_1 and T_3 have the same sign and ($|T_3| > |T_1|$), and in terms of states (incorrect) when the numerical values are difficult to deal with (Vincent, Table 4.3, for instance, in which T_1 and T_3 have opposite signs).

The functional model is not universal. When children map by forcing the problem into the functional model, they may reverse the order of transformations or even change a loss into a gain and vice versa, so that the problem suits the model.

Conne also compared T_2 problems (find T_2 knowing T_1 and T_3) and T_3 problems (find T_3) and found several interesting differences. First, the variety of answers is smaller for T_2 problems than for T_3. Second, students are more inclined to combine data and make a calculus (usually a subtraction) for T_2 problems. Finally, when students' answers consist of repeating data, they usually repeat only one datum for T_2 problems and both data for T_3 problems.

Marthe (1979) obtained more evidence in experimenting with older students (secondary school, 11-year-olds to 15-year-olds). He concentrated his attention on Category IV and V problems that can be represented by the equation $a + x = b$ where a, b, and x are directed numbers. He studied two cases: when a and b had opposite signs ($a > 0$, $b < 0$, and $|a| > |b|$), and when a and b had the same sign ($a > 0, b > 0$ and $|a| > |b|$).

His results show that the "opposite-sign" problems are always more difficult than the "same-sign" problems. For same-sign problems, Category IV was very difficult, even more difficult than the Category VI problems. There is an increase in performance across grades. However, the majority of students at the age of 15 still perform poorly. This indicates that the growth of understanding of the conceptual field of additive structures takes a long time and is not completed by the age of 15.

Experiment 3. Fisher (1979) compared different problems requiring subtraction. He worked with second-grade students (7- and 9-year-olds) during the time they learn subtraction. He used three classes of problems and different numerical data in each case.

Fisher wanted to know whether these three classes of problems were hierarchically ordered and how learning subtraction would modify students' behavior. His results showed that Category II final-state problems, in which students have to find out what the final state is knowing the initial state and the transformation are much easier than others. Performance is very low for most other problems. The order of the data is not a relevant (or important) factor. Instead, the most relevant factor is the structure of the problem.

Fisher's main conclusion is that subtraction is first understood as a direct negative transformation, not as the inverse of addition of measures, nor as the inverse of a direct positive transformation. This is very important because it has to do with the "dynamic" versus "static" problem and the relationship between addition and subtraction. If the very first model of subtraction for the child is the Category II structure, one must specify that Fisher's conclusion is true only for the "find-the-final-state" paradigm. Subtraction is not merely the inverse of addition. It has its own meaning as a direct transformation of the state of nature.

Fisher also tested the level of students on the class inclusion test and found that intermediary level 2b (just before the last level, according to Piagetian description) was a necessary condition for success on Category I subtraction problem. After six months, the order in which the ability to solve various classes of problems was acquired was in accordance with the hypothesized hierarchy. Furthermore, progress was better on Category I problems for students who correctly solved Category II final-state problems at the beginning of the school year. The most frequent error observed was giving one of the numerical data as the answer.

ARE SYMBOLIC REPRESENTATIONS USEFUL?

Although the problem of representation has been raised before in this chapter, it has not been discussed. My use of diagrams was intended to express clearly the differences my analysis takes into account. Different symbols for different sorts of addition were used in the same way. The question is: *How useful are symbolic representations?* This is not a rhetorical question. When children solve a problem, they often make the calculations first and write the symbolic representation, whatever it is, afterwards. When one studies the capacity of students to solve subtraction problems either in the form of a ready-made arrow diagram or in the form of a verbal problem, one does not find much difference. Yet, it is hard to conceive that symbolic representations are useless. Mathematicians have devoted long hard discussions to the problem of symbolization and we spontaneously use symbols to make other people understand what we are talking about. I would like to propose two criteria for the efficiency of symbolic representations.

First criterion: *Symbolic representations should help students to solve problems that they would otherwise fail to solve.*

This criterion is not easily met. We must make an effort to imagine problems for which it may be verified. My suggestion is that symbolic representations may be helpful when there are many data and when there are different structures.

Second criterion: *Symbolic representations should help students in differentiating various structures and classes of problems.*

This criterion is nearly a direct consequence of the first. Even if symbolic representations are not used to solve problems, they may be useful in helping students to analyze and differentiate structures.

Whereas the first criterion is an immediate or short-term criterion, the second one should permit a long-term evaluation. Of course, these criteria should be used to evaluate different sorts of symbolic systems, at different stages of the acquisition of additive structures. For instance, Euler–Venn diagrams may be helpful for certain classes of problems at the beginning of primary school and equations may be more helpful at the end of secondary school. A similar hypothesis may be made for arrow diagrams and for other symbolic representations.

Thus, we should take time to examine different symbolic systems and see what they can symbolize correctly, what limits they have, and what their advantages and inconveniences are. To illustrate this question let us look at two problems and five different representations for each of them (see Table 4.5).

It can be seen from Table 4.5 that Euler–Venn diagrams and distance diagrams do not discriminate problems A and B. Algebraic equations are ambiguous

TABLE 4.5
Five Symbolic Representations for Two Problems

A. Peter played marbles with friends. In the morning, he won 14 marbles. In the afternoon he lost 31 marbles. He now has 23 marbles. How many marbles did he have before playing?

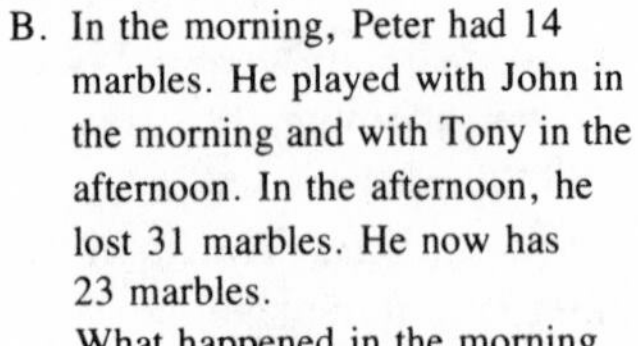
B. In the morning, Peter had 14 marbles. He played with John in the morning and with Tony in the afternoon. In the afternoon, he lost 31 marbles. He now has 23 marbles. What happened in the morning when he played with John?

1. Euler–Venn Diagrams

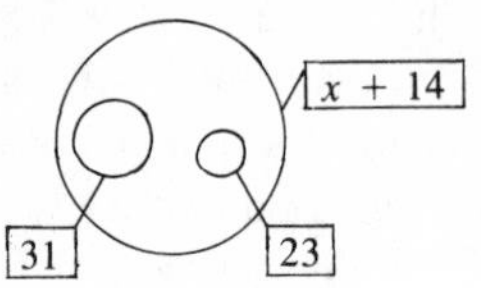

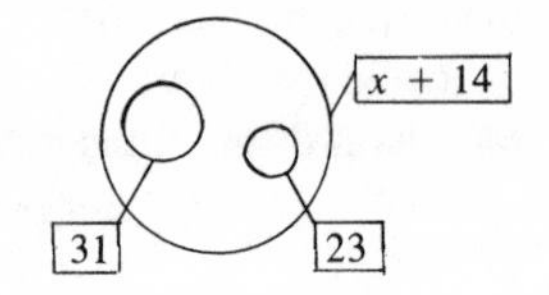

2. Transformation Diagrams

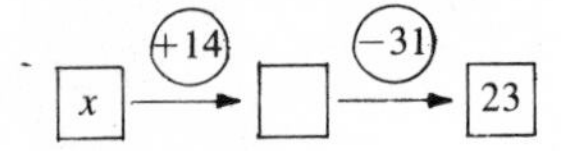

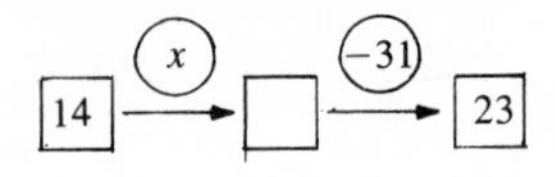

3. Algebraic Equations

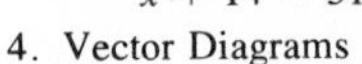
$$x + 14 - 31 = 23$$

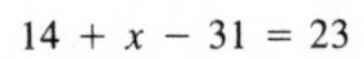
$$14 + x - 31 = 23$$

4. Vector Diagrams

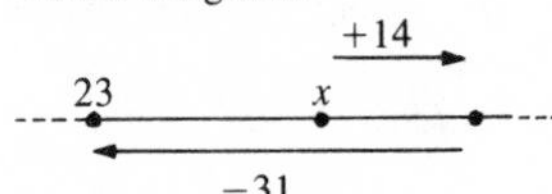

5. Distance Diagrams

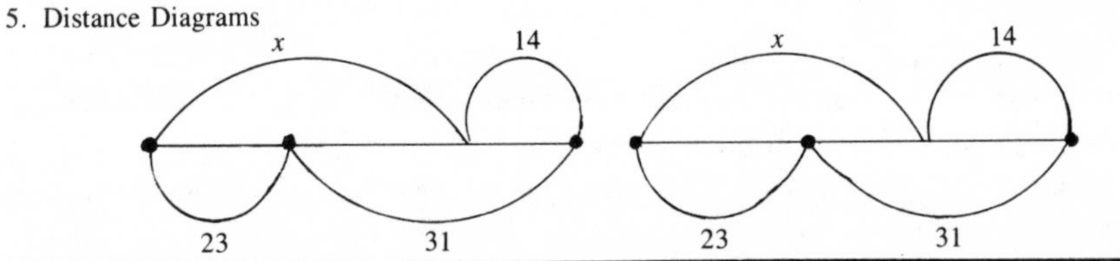

because $x + 14$ in problem A and $14 + x$ in problem B do not mean the same thing (if x were negative in problem B, it would still be written $14 + x$). Transformation diagrams and vector diagrams discriminate A and B but, before representing problem B by vectors, you have to think of the sign of the transformation x to know which way it goes. In other words, vector diagrams might suppose that you have already solved the problem. It seems to me that Euler–Venn diagrams and distance diagrams also suppose that the problem has already been solved. The only way to decide is to make experiments.

I wish I could show clear empirical results to sustain my theoretical views. This is not possible. I can mention only one experiment. I do it very briefly, because it does not concern the initial learning of additive structures. This experiment was made with 11- to 13-year-old students (last grade of primary school—first grade of secondary school) on a variety of Category I, II, III, and IV problems for which children had to deal with more than two data and more than one structure. Two versions in natural language had been written for each structure (see structures in Fig. 4.8). We compared the distributions of the procedures used by subjects for pretest and posttest sessions. Between the test sessions, students had worked 5 weeks with other problems, equations, and arrow diagrams. Table 4.6 shows results concerning the use of equations and the use of diagrams. The results can be summarized as follows:

Place of the unknown. Suppose a student writes $x =$ ____ or ____ $= x$. In these instances, the other member of the equation is usually nothing else but the sequence of numerical operations needed to calculate x. In most cases, as in the five structures, the unknown x should appear as an element inside one member of the equation.

Chaining equalities. Fairly often, students are able to make a correct response but they write a wrong statement because they chain equalities that should be kept separate. For instance,

Step 1: $1063 + 217 = 1280$.
Step 2: Instead of writing on the next line
$1280 - 425 = 855$,
they pursue on the same line
$1063 + 217 = 1280 - 425 = 855$.

This error clearly indicates that for many students the equality sign does not stand for a relationship between numbers but for a procedure to calculate the unknown.

Inverse writing of subtraction. Writing the sentence $425 - 1280 = 855$ instead of $1280 - 425 = 855$, is a well-known error. It means that the student is not interested in the sentence, only in the result. The sign "−" represents the

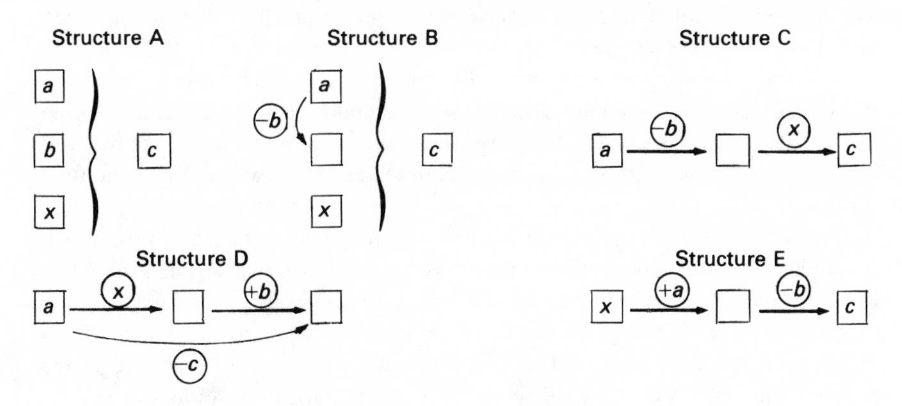

FIG. 4.8. Five structures used to study use of diagrams.

TABLE 4.6
The Use of Equations and Diagrams on a Set of Problems

Equations

Unknown placed inside an expression

	CM_2	6 éme	Total
Pretest	1	1	2
Posttest	8	8	16

Chaining equalities (error)

	CM_2	6 éme	Total
Pretest	7	7	14
Posttest	7	6	13

Inverse writing of subtraction (error)

	CM_2	6 éme	Total
Pretest	19	5	24
Posttest	8	12	20

Diagrams

Correct use of diagrams

	CM_2	6 éme	Total
Pretest	1	0	1
Posttest	31	8	39

CM_2: 50 subjects–last grade of elementary school (11 years)
6 éme: 50 subjects–first grade of secondary school (12 years)
Total: 100 subjects

difference between two numbers, regardless of which the big one and the small one are. What matters to the student is not the problem of writing a correct relationship but the problem of finding the result.

Both chaining equalities and inverse writing of subtraction are understandable. However, they undoubtedly mean that equations and equalities are not used to extract and represent relevant relationships between numbers but rather to recall the sequence of numerical operations used to calculate a result. Table 4.6 shows that only a few students (2% at the pretest and 16% at the posttest) are able to place the unknown inside an expression at least once in the test. It shows also that the two classical errors (chaining equalities and inverse writing of subtraction) are frequent.

Diagrams can be considered a kind of equation with additional information specified (measures, states, transformations, or relationships). So the problem is to determine whether they can be used more easily by children to represent problems. Table 4.6 shows that although students were not acquainted with diagrams at the pretest, almost 40% of them were able to use diagrams at the posttest. This makes credible the thesis that, at this stage of development (11–12 years), diagrams are more appropriate than equations. But we still have to prove it.

CONCLUSION

In the last part of this chapter, the term *representation* has been used with the restrictive meaning of ''symbolic system'': a set of signs, syntax, or operations on the elements of the system. Earlier I referred to different categories that are essentially conceptual. Concepts and symbols are two sides of the same coin and one should always take care to view students' use of symbols in the light of their use of concepts. In other words, the ability to solve problems in natural language, issued from ordinary social, technical, or economical life, is the best criterion of the acquisition of concepts. Reciprocally, it is essential to know how mathematical symbolization helps students. My view is that the acquisition process consists of building up relational invariants, analyzing their properties, and building up new relational invariants. Some quantitative invariants have been studied by Piaget within the context of conservation experiments. Little work has been done on relational invariants. We should do more, because arithmetic, geometry, physics, and other fields of knowledge consist essentially of relational invariants.

In this chapter, I have dealt with relationships that involve time and have shown that time is the source of important differences among the problems children try to solve. Another source of differences is the inclusion relationship. Carpenter and Moser (1979) have raised this problem under the ''part–part–whole'' category, and I have also mentioned briefly some results, obtained by Fisher (1979), that show the relevance of this aspect.

Looking at the three main concepts I have used in my classification (measure, time transformation, and static relationship), and at Categories I, II, and III, one can see that there are three criteria involved in these distinctions. These criteria are not independent from one another, although each of them carries some separate information.

> The first criterion is: Either all elements are measures (Category I) or one element is not a measure (Categories II and III).
> The second criterion is: Either time is involved (Category II) or time is not involved (Categories I and III).
> The third criterion is: Either there is an inclusion relationship (Categories I and II) or there is no inclusion relationship (Category III).

I have not emphasized this last criterion in this chapter. It is probably very important because Categories I and II both convey an inclusion relationship. In Category I, the elementary sets (or magnitudes) are parts of the whole; in Category II, either the initial set (or magnitude) is part of the final one, or the final one is part of the initial one. In Category III, because the two linked measures are simultaneously present, there is not necessarily (and usually there is not) any inclusion relationship. For example in "Peter has five more marbles than John," none of the quantities involved (Peter's marbles, John's marbles, or the difference) is included in any other set. Little experimental work has been done on these aspects, but I would expect it to be fruitful.

Additive structures are a difficult conceptual field, more difficult than most mathematics teachers expect. Understanding additive structures is a long-term process that starts with some simple find-the-final-state problems and goes on into adolescence with subtraction of opposite-sign transformations.

Some problems are more easily solved than others; likewise, children find some procedures more natural than others. We should use these procedures in teaching, even if they are not canonical but only local procedures, because they are probably the best way to help children build up the canonical ones. Brousseau (1978) has developed a systematic theory of how to get children to a new step, which concerns situations in which children have to act, to formulate, and to validate. For instance, changing the numerical values is very often necessary to get children to move from a local procedure to a more powerful canonical one.

What is true for problems and procedures is also true for symbolic representations; some representations such as equations are more powerful than arrow diagrams or Euler–Venn diagrams. However, these equations should represent meaningful situations. Consequently, the problem of knowing which intermediate representations may help students to handle these situations is very important. Probably equations using directed numbers are too abstract, having too many different meanings mapped onto the same sign. For instance, the minus sign may stand for a direct negative transformation, a complement, the inverse of a positive transformation or relationship, a subtraction of two transformations, and so on. Equations using natural numbers are undoubtedly easier, but either they stand for the procedure used by students and not for the objective structure of problems, or they represent correctly the structure of only a few classes of problems.

My last comments are on the concept of "theorem in action." In Piaget's experiments, obviousness is the strongest criterion of cognitive appropriation. When children do not conserve, they find it obvious that things change; when they do conserve they find it obvious that things do not change. Similarly, when children solve an additive problem, they usually find it obvious that they should add or subtract. The relational calculus they have performed, implicitly, makes it clear that they should do a particular numerical operation. This is what I call a "theorem in action." In more sophisticated terms, understanding a new relational invariant or a new property of an invariant provides the choice of the right numerical operation. The child is now able to "arithmetize" a qualitative structure.

In conclusion, I believe that didactic and psychological research in mathematics teaching should pay more attention to the following four questions:

1. What are the easiest "theorems in action" used by students in solving verbal problems?
2. How should we get students to build up new theorems by presenting them with new situations?
3. How and with the help of which symbolic systems should we help students make these theorems explicit?
4. How can we make sure that theoretical theorems actually become theorems in action?

Practice and theory are sides of the same coin, and problem solving is certainly the source and the criterion of appropriated knowledge. The natural order of appropriation, when it exists and when it can be found, will be of great value in the classroom.

REFERENCES

Brousseau, G. L'observation des activités didactiques. *La Revue Française de Pédagogie,* 1978, *45,* 130–139.

Carpenter, T. P., & Moser, J. M. The development of addition and subtraction concepts in young children. *Proceedings of the Annual Meeting of the International Group for the Psychology of Mathematical Education.* Warwick, England 1979.

Conne, F. *Pierre, Bertrand, Claude, Paul, Laurent, Michel et leurs billes. Contribution à l'analyse d'activités mathématiques en situation. Approches en psychopédagogie des mathématiques.* Université de Genève. Faculté de Psychologie et des Sciences de l'Education, Genève, 1979, 25–84.

Fisher, J. P. *La perception des problèmes soustractifs aux débuts de l'apprentissage de la soustraction.* Thèse de 3ème cycle. IREM de Lorraine, Université de Nancy I, Nancy, 1979.

Marthe, P. Additive problems and directed numbers. *Proceedings of the Annual Meeting of the International Group for the Psychology of Mathematical Education,* Warwick England, 1979.

Vergnaud, G., & Durand, C. Structures additives et complexité psychogénétique. *La Revue Française de Pédagogie,* 1976, *36,* 28–43.

5 Interpretations of Number Operations and Symbolic Representations of Addition and Subtraction

J. Fred Weaver
University of Wisconsin–Madison

This chapter is presented in two principal sections. The first section emphasizes an all-too-frequently neglected interpretation of number operations that, I believe, has considerable import for the future research and instructional programs detailed in the second section. Although this first section is relatively unsophisticated, it goes well beyond the initial learning of addition and subtraction skills.

SOME BACKGROUND CONSIDERATIONS

In many textbooks for elementary school students, no attempt is made to give explicit verbal or symbolic formulation to operations in general, and to addition and subtraction in particular. This may be a fortuitous situation, as other texts offer characterizations of the addition and subtraction operations that tend to be confusing, vacuous, or even potentially erroneous when viewed in light of more advanced interpretations. We find, for example, that operations are thought of as "rules" (Milton & Leo, 1977), as "combining" (Eicholz, O'Daffer, & Fleenor, 1978), or as a "way of thinking" (School Mathematics Study Group, 1965). For secondary students it is common to equate the notion of an operation with that of a function f where a function is defined as "a set of ordered pairs (x, y) where: (1) x is an element of the set X; (2) y is an element of the set Y; and (3) no two pairs in f have the same first element" (Allendoerfer & Oakley, 1963). Nothing in the definition precludes the possibility that $Y = X$, or that X is itself a product set.

Because of later considerations in this chapter, it is helpful to distinguish between certain kinds of operations. An n-*ary operation* on a set A is a function from $A \times A \times \cdots \times A$ (n factors) into a set A. This chapter is concerned primarily with unary ($n = 1$) and binary ($n = 2$) operations. I adhere to the characterization of an operation as a set of ordered pairs in which the assignment of the second member to the first may be made arbitrarily but more often is done so according to some rule. Because younger children are the major interest of this volume, I deal with the addition and subtraction of whole numbers, for which the following characterizations may be used:

An assignment rule for addition of whole numbers:

> Select sets A and B such that $A \cap B = \phi$, $n(A) = a$ and $n(B) = b$. Then $c = n(A \cup B) = a + b$ where $A \cup B = \{x \mid x \in A \text{ or } x \in B\}$.

An assignment rule for subtraction of whole numbers:

> Select sets A and B such that $B \subseteq A$, $n(A) = a$ and $n(B) = b$. Then $c = n(A \setminus B) = a - b$ where $A \setminus B = \{x \mid x \in A \text{ and } x \notin B\}$.

Strictly speaking, addition is a binary operation on the whole numbers but subtraction is not because, for certain pairs of numbers, it is impossible to subtract and have a whole-number answer (e.g., $7 - 13$). However, in this chapter it will be easier to speak of both addition and subtraction as binary operations, recognizing that subtraction is limited to those number pairs for which it is possible to carry out the operation and obtain a whole number result.

It is well to note that the foregoing rule is not the only way to describe addition or subtraction. The same operation of addition, for example, could be derived from a concatenated segments assignment rule:

> Select distinct collinear points X, Y, Z such that Y is between X and Z, $m(\overline{XY}) = a$, and $m(\overline{YZ}) = b$. Then $m(\overline{XZ}) = c = a + b$.

It is also extremely important to recognize that regardless of the assignment rule associated with addition, subtraction can be characterized directly in terms of the addition operation in this way:

$$a - b = c \text{ means there exists a number } c \text{ such that } c + b = a.$$

Certain mathematics texts at the teacher level use this interpretation of subtraction, establishing the fact that $a - b = n \leftrightarrow n + b = a$ as the basic way of defining subtraction, although they may not make explicit that $a - b = n \leftrightarrow b + n = a$. Other texts choose the latter as the definition and may not make the former explicit. For adult comprehension, the distinction may be trivial. However, for young children in their development of ideas about addition and subtraction, this distinction may be highly significant.

Unary Operations

Turning now to a different conceptual matter, consider the expression

$$a \text{ o } b = c$$

where a, b, and c are members of a set S and o stands for the binary operation that assigns to the ordered pair (a, b) in $S \times S$ a unique image c in S. It is also possible to interpret this expression as a unary operation. A post- or right-*operator* "o b" signifies a unary operation that assigns to operand a in S a unique image c in S. (Where needed to make this explicit, the notation $a \, \underline{\text{o}\, b} = c$ will be used.) It is also possible to think of a pre- or left-operator but because this presents distinct difficulties when the operation is subtraction, it will not be considered further in this chapter. In connection with the notion of a unary operator, it is imperative that symbols such as "$+b$," "$-b$," "$+7$," or "-5" be interpreted as unary operators and not as signed numbers, directed segments, or vectors.

There are distinctions to be made between binary and unary operations. The first deals with the commutativity of addition, which states that for all numbers a and b:

$$a + b = b + a$$

Within the whole-number domain, it is also valid to assert that

$$a \, \underline{+ b} = b \, \underline{+ a} \text{ or, in particular, } 7 \, \underline{+ 2} = 2 \, \underline{+ 7}.$$

On the surface, these latter expressions appear to exemplify commutativity, but really they do not. Except when $a = b$, the unary operators "$+a$" and "$+b$" signify different operations; for instance, "$+2$" and "$+7$" are not the same operation. This "pseudocommutativity," as I choose to term it, is a valid property, but not about an operation. In the following cases the unary addition and subtraction operators do commute:

$$9 \underline{+6} \underline{+3} = 9 \underline{+3} \underline{+6} \quad 9 \underline{-6} \underline{-2} = 9 \underline{-2} \underline{-6} \quad 9 \underline{-6} \underline{+2} = 9 \underline{+2} \underline{-6};$$

but in this case they would not:

$$5 \, \underline{+ 2} \, \underline{- 6} \neq 5 \, \underline{- 6} \, \underline{+ 2}$$

because on the right-hand side, the initial operand 5 is of insufficient size to permit the first unary operation of "-6."

A second distinction deals with the erroneous assertion that binary addition and subtraction are inverse processes. Because a binary operation is a rule or mapping that assigns a single unique number to an ordered pair of numbers, it is not possible to reverse the process and assign an ordered pair to a single number. On the other hand, infinitely many pairs of unary operations are inverses of each other. It is quite correct to make statements such as:

$$a \underline{+b} \, \underline{-b} = a \qquad a \, \underline{-b} \, \underline{+b} = a \qquad 7 \, \underline{+2} \, \underline{-2} = 7 \qquad 7 \, \underline{-2} \, \underline{+2} = 7$$

provided that a is a proper operand in the sense that the unary operation of "$-b$" is possible.

The third distinction between binary and unary interpretations of addition and subtraction relates to the notion of these operations making something more or less. To think of comparing magnitudes with respect to the binary interpretation seems senseless because it cannot be asserted that $c > (a, b)$ or $c < (a, b)$. On the other hand, for the unary operation, it does indeed make sense to think of addition as making more and subtraction as making less within the domain of the whole numbers, provided the unary operation is nonzero. In the next section I consider these "change-of-state" interpretations of unary operators in more depth.

Some Instructional and Research Considerations

A major aim of arithmetic instruction for children is to characterize a symbol system they can use to represent events, actions, and relationships with which they have had direct experience. At the elementary levels, these experiences are primarily the manipulation of objects. An important part of the symbol system is the open and closed addition sentences of the form $a \pm b = c$, where, for open sentences, the letter n or x or a □ may be used in either the "a," "b" or "c" position to represent a missing number. There is a principal domain of embodiments to which appropriate actions are applied when referents for the meaning or interpretation of those symbolic sentences are being presented. Although verbalizations often accompany tasks involving these embodiments, such verbalizations should not be confused with verbal problems to which number operations may be applied. Verbal problems do not serve as a referent for symbolic sentences in any direct sense although they may do so, indirectly, when they generate actual or imagined embodiments associated with some symbolic number sentences.

Binary or Unary: A Research Question

As indicated in the first section, exemplars of number sentences are subject to the ambiguity of a binary operation interpretation versus a unary-operator interpretation. This is further confounded by referent differences that may be encountered in connection with various embodiments. Regardless of the binary or unary interpretations, discrete versus continuous embodiments could have a nontrivial bearing on meanings children attach to symbolic number sentences. For instance, combining two sets of cardinality 4 and cardinality 2 to form a single set of cardinality 6 ($4 + 2 = 6$) is conceptually different from starting with a set of cardinality 4 and joining to it a set of cardinality 2 to form a set of cardinality 6 ($4 \, \underline{+2} = 6$); in either case, working with sets of cardinalities 4, 2, 6 is different from working with analogous combining or joining actions on representations of line segments having measures of 4, 2, 6.

It is highly desirable, and possibly even necessary, that at some time pupils acquire both binary-operation and unary-operator meanings for symbolic sentences. Yet there now seems to be little or no direct research evidence regarding the answers to the following questions:

> How, in fact, do young children interpret any of the symbolic number sentence types?
> Do, or can, binary and unary interpretations develop more or less simultaneously among young children?
> If not, which interpretation should (or does) come first and when should (or does) the other follow?
> Under what condition, if any, is interference between the two interpretations likely?

On the other hand, researchers studying children's solution procedures with respect to verbal problems have identified a variable not unrelated to interpretations and meanings associated with symbolic sentences. This is a change-of-state situation called by a variety of names. (In addition to the chapters in this volume by Carpenter & Moser, Nesher, and Vergnaud, see also Blume [1981], Greeno [1979], and Steffe & Johnson [1971]). This change of state is a situation of action, in contrast to static situations as in comparison or part–part–whole relationships. In none of the works cited here, however, have these change-of-state situations been related in any explicit way to the distinction between unary and binary operations on numbers.

Fuson's (1979) analysis and model of verbal problem situations explicitly relates change-of-state and other contexts to a unary–binary distinction. Fuson suggests that her classification of verbal problem situations might be developmentally ordered with change-of-state situations associated with unary operations having primacy. She contends that support for this hypothesis comes from several sources (Brush, 1972, 1978; Gelman & Gallistel, 1978). In each case young children worked with change-of-state embodiment tasks and conclusions were reached about children's preinstructional conceptualizations of addition and subtraction (in reality, children's ideas about unary operators, although not identified as such by the investigators). Understandably, at these preschool levels no attempt was made to associate embodiment tasks with symbolic number sentences.

A Needed Research Direction

Several research programs are under way that focus attention principally upon verbal problems. But there is need for further investigations of children's interpretation of various open sentences. A recent investigation by Lindvall and

Ibarra (1980) confirms suspicions from previous studies that toward the end of grade 1 and the beginning of grade 2, pupils' comprehension of the meaning of noncanonical open sentences is far from clear.

The kind of research I believe is needed might be termed instructional, investigating the effect of direct instruction within classroom contexts.

The intent is, through instruction, to develop unary-operator interpretations for various symbolic sentence types. The unary-operator approach means that sentences are to be interpreted in accord with this generic form, which is intended to encompass all sentence types:

Initial state	Change of state (Unary operator)		Final state
a	$\underline{\nabla\ b}$	$=$	c

Finding an unknown change of state requires that two things be specified, whereas finding an unknown final state or initial state requires that only one thing be specified: a number.

Interpretation and meaning for number sentences will be derived principally from appropriate instructional tasks involving concrete, pictorial, or disgrammatic embodiments. Symbolic mathematical referents also have some contribution to make since, for instance, one interpretation of exemplars of the form $a - b = c$ may be derived from exemplars of the form $c + b = a$.

It is hypothesized that there is a distinct advantage in using a mediating notational form prior to the introduction of conventional notation (and possibly for an extended period of time). Support for this hypothesis comes from Fuson (1979) and Vergnaud (see Chapter 4 in this volume). One candidate for a mediating notational form is the common and familiar arrow diagram: notation; $a \xrightarrow{+b} c$. Another possible candidate, a bit more cumbersome than arrow diagrams, is the flow chart (Fig. 5.1). In any event, the use of some preliminary or mediating

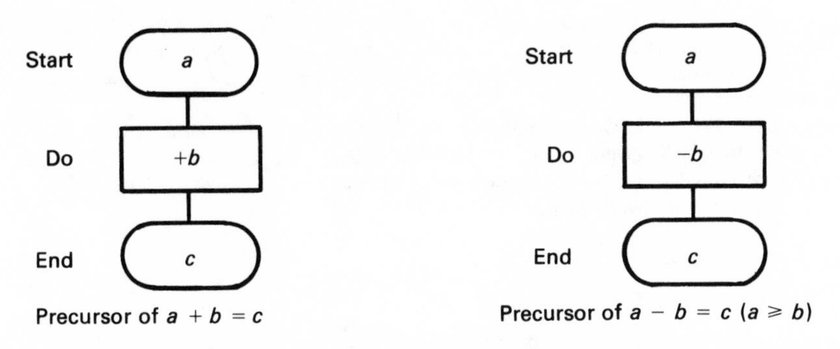

FIG. 5.1. Mediating notational form, the flow chart.

forms prior to conventional number-sentence or equation notation should be investigated.

Obviously, these suggestions for research raise many questions, including the role of verbal problems, which I have not discussed. I believe, however, that the development of students' ability to attach valid interpretations and meanings to symbolic sentences begins with change-of-state situations and embodiments in the context of instructional tasks. I believe these situations may be interpreted as unary operations and operators, with children being provided some mediating notational forms at first instead of conventional mathematical symbols.

REFERENCES

Allendoerfer, C. B., & Oakley, C. D. *Principles of mathematics.* (2nd ed.) New York: McGraw-Hill, 1963.

Blume, G. W. *Kindergarten and first-grade children's strategies for solving addition and subtraction problems in abstract and verbal problem contexts.* Unpublished doctoral dissertation. The University of Wisconsin–Madison, 1981.

Brush, L. R. Children's conception of addition and subtraction: The relation of formal and informal notions (Doctoral dissertation, Cornell University, 1972). *Dissertation Abstracts International,* 1973, *33,* 4989B. (University Microfilms No. 73-10-100).

Brush, L. R. Preschool children's knowledge of addition and subtraction. *Journal for Research in Mathematics Education,* 1978, *9,* 44-54.

Eicholz, R. E., O'Daffer, P. G., & Fleenor, C. R. *Mathematics in our world* (Levels 11-16; Grade 3). Menlo Park, Calif.: Addison-Wesley, 1978.

Fuson, K. C. Counting solution procedures in addition and subtraction. Paper presented for the Wingspread Seminar on the Initial Learning of Addition and Subtraction Skills, Racine, Wis., November 1979.

Gelman, R., & Gallistel, C. R. *The child's understanding of number.* Cambridge, Mass.: Harvard University Press, 1978.

Greeno, J. G. Preliminary steps toward a cognitive model of learning primary mathematics. In K. C. Fuson & W. E. Geeslin (Eds.), *Explorations in the modeling of the learning of mathematics.* Columbus, Ohio: ERIC Clearinghouse for Science, Mathematics, and Environmental Education, 1979.

Lindvall, C. M., & Ibarra, C. G. Incorrect procedures used by primary grade pupils in solving open addition and subtraction sentences. *Journal for Research in Mathematics Education,* 1980, *11,* 50-62.

Milton, K., & Leo, T. J. *Active interest math* (Enrichment, Card 4). Newton, Mass.: Selective Educational Equipment, 1977. (Originally published in the UK by Creative Educational Press PTY, 1975).

School Mathematics Study Group. *Mathematics for the elementary school, Grade 4: Student's text, Part I* (Rev. ed.). Stanford, Calif.: Stanford University, 1965.

Steffe, L. P., & Johnson, D. C. Problem-solving performance of first-grade children. *Journal for Research in Mathematics Education,* 1971, *2,* 50-64.

6 An Analysis of the Counting-On Solution Procedure in Addition

Karen C. Fuson
Northwestern University

This chapter presents a structural analysis of the counting-on procedure in children's addition and offers data pertaining to that analysis. Particular attention is given to the use of the sequence of counting words (one, two, three, etc.) as a representational tool for counting on. Because little detailed empirical or theoretical work has addressed this area until recently, the analysis proposed here will undoubtedly require modification and elaboration as additional evidence is generated.

COUNTING ALL AND COUNTING ON

The counting-on procedure derives from a simpler counting solution procedure called counting all. Both procedures involve the representation of the addends by entities and the subsequent enumeration of the entities for at least one of the addends. In counting all, the sum is determined by a count of the total number of entities comprising the two addends; the final verbalization for the problem "five plus three" would be "one, two, three, four, five, six, seven, eight; the answer is eight." In counting on, the enumeration begins at the word for the first addend and continues until the entities representing the second addend have been enumerated; the final counting-on verbalization for the same problem would be "five, six, seven, eight; the answer is eight." An even more efficient counting-on process is to begin with the larger of the two addends. For the sake of simplicity, the counting-on process is described as if it always begins with the first addend, but children who are counting on often use the more efficient procedure.

Addition problems can be presented in various ways, and aspects of both counting all and counting on may vary with the mode of presentation of the problem and with the particular solution aids available. The counting-all procedure requires the use of entities to be counted. These may be provided as the problem itself (e.g., "How many cookies are here?") or as supplements to a verbal or symbolic problem, or children may have to generate the sets for each addend from available countable objects (chips, fingers, dots on paper, a mental representation, etc.). This paper will not deal with the simplest of addition problems, the presentation of two sets of objects with the question "How many in all?" When an addition problem situation is referred to in this paper, it is assumed that the number for each addend is provided verbally or symbolically.

The counting-on procedure is obviously more efficient than the counting-all procedure, at least where large sets are concerned, but its use requires some fairly sophisticated understandings. Some of these concern the first addend, some concern the second addend, and others concern the correct connection of the two addends. Each of these aspects of counting on will be discussed in turn.

THE STRUCTURE OF THE FIRST ADDEND IN THE COUNTING-ON PROCEDURE

In the counting-on procedure the final counting of entities for the sum begins with a single word for the first addend. We argue that this first single word produced in counting on carries with it four different, though related, meanings, and the confluence of all these different meanings permits the counting-on procedure to function. The view that number words carry different meanings, dependent on the context in which they are used, is discussed in more detail in Fuson and Hall (in press), where literature about early number concepts and counting is also reviewed.

The first two meanings of the first counting-on word that must be related are a cardinal meaning and a counting meaning. The transfer from the counting meaning to the cardinal meaning has been identified as the cardinality rule (Schaeffer, Eggleston, & Scott, 1974) or the cardinal principle (Gelman & Gallistel, 1978). This rule states that the last counting word produced in the enumeration of a set of objects is the cardinal number for that set. To know how many objects there are in some set, a child must therefore connect the counting meaning of the last word produced with the cardinal meaning for that word. In counting on, the dual-meaning relationship also must be applied in the opposite direction. The child is told the cardinal number of the first addend (e.g., "there are nine squares here") and must know that if the set of objects for the first addend are counted, the last counting word produced will be nine. This bidirectionality of the dual meaning of the first counting-on word can be seen more clearly as follows:

1. The count-to-cardinal direction (cardinal principle) summarizes an action that has been completed.
2. the cardinal-to-count direction predicts an action that may be carried out.

The mastery of the dual meaning in the first direction underlies counting all. Children must master the second direction before they can count on. The second direction also is required if a child is making a set of objects to count for either the counting-all or counting-on procedure.

Once the cardinal-to-count connection has been made (i.e., given "nine" a child knows the count will be "one, two, . . . , nine"), the actual production of the count itself can be abbreviated to the simple production of the final counting word "nine." This is the third meaning of the first word produced in counting on: The word is also an abbreviation of the whole counting segment for the first addend. In other words, "nine" also is an abbrevition of "one, two, . . ., nine." Thus, the production of the single word "nine" is a summary of the verbal enumeration of the first addend and is substituted for this enumeration.

The first three meanings of the first addend word all concern the addend per se. The fourth meaning relates the first addend to the second addend within the context of the final counting-all sequence. The enumeration of entities for the first addend (e.g., for 9 + 5, "one, two, three, . . . , nine") occurs both alone and as the first part of the enumeration of the entities in both of the addends ("one, two, . . . , fourteen"). When the first addend enumeration is able to be seen simultaneously as an enumeration of the first addend and as a part of the count-all enumeration of the sum, the abbreviation of that enumeration (the word "nine") can be substituted for the one-to-nine enumeration *within* the whole enumeration of the sum set. This substitution is what leads to the counting-on procedure: the production of the first addend word as the abbreviation of the first addend enumeration and the continuation of the enumeration of the entities in the second addend ("nine, ten, eleven, twelve, thirteen, fourteen"). The third abbreviation meaning of the first counting-on word is a summary of the counting of the first addend; the fourth meaning is the starting point for the continued enumeration of the second addend. Steffe, Thompson, and Richards (this volume) term this aspect of counting on "extension."

This fourth meaning arises from the fact that each addend plays a double role: It is both an addend and a part of the sum. This double role is played either by a sequence of counting words (e.g., "one, . . . , nine") as previously described or by a set of entities representing the first addend. The entities or the words initially represent the first addend and then represent the first part of the sum. It seems possible that children might discover this double role of the first addend by focusing either on the objects ("I am just counting these same objects again. I don't have to do that.") or on the verbal enumeration itself ("I'm just saying these same words again. I don't need to do that"). Focusing on the words themselves might facilitate the substitution of the abbreviation of the word "nine" for the whole enumeration of the first addend, but it would seem to reduce the salience of the cardinal meanings involved in the addition problem. If the cardinal aspect is totally missing, a type of error reported in the third section of this chapter, which indicates an incorrect connection of the first and second addends, seems more likely than when children construct counting on by focusing on objects. The possibility that children do arrive at counting on by these two different paths, with the consequent differences in susceptibility to certain types of errors, would seem worthy of exploration in future research. The available data address this issue only indirectly.

A final point needs to be made with respect to the first addend word. In some circumstances or in some people the production of the first addend word becomes abbreviated, and overt counting on begins with the enumeration of the second addend. We have observed a few children who would do a problem such as 14 + 6 by counting "fifteen, sixteen, seventeen, eighteen, nineteen, twenty." Counting on in this way would decrease the likelihood of the addend connecting error described later, but the complexity of meanings involved in the first addend word indicate that it would be helpful, at least initially, to express this word overtly.

A Study of Addition Solution Procedures

The foregoing analysis was based both on informal observations of children and on problems adults displayed in trying to count on using a sequence of words that could be used in an enumeration procedure but that possessed no cardinal meanings—the alphabet. A study with children was undertaken to provide some preliminary data about behaviors concerning the four meanings of the first counting-on word discussed previously and about conditions under which counting on will spontaneously occur (Secada & Fuson, in preparation). In this study relatively large numbers (12 through 18) were used for the first addend and the numbers 6 through 9 were used for the second addend in an attempt to avoid the children's use of both known addition facts and special solution procedures applicable only to very small subitizable numbers (four or under). The presence of objects was manipulated in order to examine the effect of object presence on solution procedure. This study was designed to provide data concerning both the first and the second addend in the counting-on procedure.

Twenty-eight children aged 6 to 8 from an educational demonstration school in a predominantly white middle- to upper middle-class suburb served as subjects. Each subject was presented eight addition problems adapted from Steffe, Spikes, and Hirstein (1976).

Each addend was presented in two ways: by a numeral written on a small card and by a row of 1-inch squares drawn on a separate longer card. Two conditions of card arrangement were used. In the first condition *both* cards containing the rows of squares were faceup and therefore visible to the child. In the second condition, only the card containing the row of squares for the *second* addend was visible; the card with squares for the first addend was turned facedown so that the squares were not visible to the child. The numeral for each addend was visible at all times in both conditions. Each child was given both conditions in counterbalanced order. Only one child was unable to solve any of the problems. This child, who was in kindergarten, could not read the numerals properly and was dropped from the study.

Results. Five of the 27 subjects exhibited procedures transitional between counting all and counting on. Their behavior seemed to indicate that the passage to counting on occurs when a child withdraws attention from the entities during an enumeration of the total sum and focuses on the words themselves. The word production was not yet counting on, however, because these children always started with "one." Two of the five children produced counting words for the first addend when entities for the first addend were visible, but the counting run of the verbal string was very quick and careless compared to the slower object-focused counting of the second number. The other three children counted all when the first addend squares were visible but exhibited this transitional process when the first addend squares were not visible. They produced counting words up to the first addend and then continued counting the squares for the second addend. Thus these five subjects seemed to need to run through the counting words for the first addend even though this run-through was not part of an actual enumeration of objects in the final count-all step. This drive to say the whole count segment for the first addend was particularly striking because the numbers in this study were relatively large; it took some time and effort for the child to run off a count to 12, 16, or 18. The existence of this transitional word sequence behavior seems to indicate some difficulty with one of the last two meanings of the first count-on word: the abbreviation meaning or the coordination of the first addend and final count meanings.

In contrast to the word focus of these transitional children, the two subjects who persisted in trying to count all when the first addend entities were no longer visible remained focused on the entities rather than on the words or the first addend numeral. They counted while they pointed at the facedown card, moving their finger along the card as they counted as if they were counting a row of squares. Thus, they seemed to be counting a *figural representation* that they generated for themselves on the blank card. For one child, the generated figural representation always took precedence over the information on the numeral card. When this child reached the end of the first array card while doing this approximate counting procedure, he continued counting the squares on the second card even though the last count word produced for the blank card did not coincide with the numeral on the digit card. This same pattern was observed for the other child for the first two trials, but on the third and fourth trial this child pointed several times near the end of the first blank card, seemingly waiting until the count sequence reached the displayed numeral for the first addend before continuing on to count the objects on the second card. Thus the second, but not the first child gave evidence of understanding the cardinal as well as the count meaning of the first addend word.

Five children used both the counting-all and counting-on procedures. They used counting on in the conditions in which entities were not present for the first addend but used counting all when squares were visible for both addends, even when the latter condition came second. Thus the presence of all the squares caused them to use a less efficient procedure than they were capable of using. Carpenter and Moser (this volume) and Davydov and Andronov (1981) also reported this type of regression. This finding indicates that the acquisition of counting on is not an all-or-nothing phenomenon and that despite the extra time and effort required for counting all, children will at least for a time choose to count all when they can do so even when they also can count on.

Parts of our analysis of the meanings attached to the first addend are remarkably similar to that in a paper by Davydov and Andronov (1981) who argue that adding on (their term for counting on) requires simultaneous consideration by the child of the "ordinal" and "quantitative" aspects of the number word (our counting and cardinal aspects). They differentiate formal counting on from actual counting on. Formal counters-on point to a single element of the first set as they state the first addend word and, if pushed by the experimenter asking questions such as "Is this really four? Why, it is only one," will point to a different object, quit trying, or count all. They comprehend only the counting meaning of counting on. Actual counters-on say the first addend number in a drawled out manner while making a sweeping gesture over the objects representing the first addend and then continue counting (e.g., "s - i - x, seven, eight"). Davydov and Andronov argue that the sweeping gesture indicates the cardinality of the first addend and thus that these children understand both the counting and cardinal meanings.

In our study the behavior of the children who counted on in either condition was examined for any indication of the meanings a child attached to the first addend word. Twelve of the 13 children who counted on when entities for both addend were visible and 14 of the 16 subjects who counted on when only the second addend entities were visible made no gesture of any kind when saying the number word for the first addend. The word was simply stated, usually in a slow emphasized way, and then the entities of the second set were counted beginning with the next count word. The gestures made by the other children (e.g., pointing at the end of the first addend card) seemed to indicate a counting meaning attached to the last square in the first addend. Only one subject on one trial in our sample made the sweeping gesture described by Davydov and Andronov. Because on a later trial this subject pointed at the end of the facedown card where the last square would have been, his sweeping gesture would seem to have referred to an abbreviated act of enumeration rather than to the cardinality of the first addend. Davydov and Andronov report that the sweeping gesture later becomes abbreviated and disappears. Because the children in our study were not probed in the same way, it is not clear which of them were formal and which were actual counters-on. The failure to find the sweeping gesture in this sample may indicate that it is a very transitory phenomenon or that it may not be used by all children.

Davydov and Andronov reported two studies in which they attempted to train counting on. The unsuccessful method focused on two counting meanings (one with objects and one with words), whereas the successful one linked a counting and a cardinal meaning for the first addend.

Davydov and Andronov also made a rather nice distinction between counting on and what we might term adding on (i.e., between mental or symbolic addition and practical object-oriented addition). They gave certain problems in which the number of objects in the first set conflicted with the number of objects the child was told that set contained. When children were able to use the number of objects they were given verbally rather than the number given perceptually, Davydov and Andronov concluded that they had constructed the mental "action of addition as distinct from the action of counting." At both levels the cardinal meaning and the counting meaning are related, but only with the latter mental action of addition does the cardinal meaning of a given spoken word predominate over the perceptual-counting meaning of a conflicting set of objects. This distinction might result as a child moved from our first meaning, a counting word for the last entity enumerated in the first addend, to our third meaning, the word as an abbreviation of the first addend enumeration. To the extent that the word itself is such an abbreviation of that enumeration, the enumeration of the entities is superfluous.

The current state of the evidence seems to indicate that a child may progressively move from a focus on our first meaning to the second and then to the third of our meanings of the first addend word.

THE STRUCTURE OF THE SECOND ADDEND IN COUNTING ON

In counting on, the counting words produced for the second addend do not coincide with the words that would be produced in counting the second addend by itself. When the second addend is small (up to about four), even 6-year-olds seem to be able to start at a given word and produce the next words without special keeping-track methods (see the counting up/counting down study in Fuson, Richards, & Briars, in press). However, when the second addend is larger than four, some means must be used to keep track of how many words have been produced beyond the first addend.

We have explored keeping-track processes in three studies with adults and two with children. We have classified the methods spontaneously employed by these

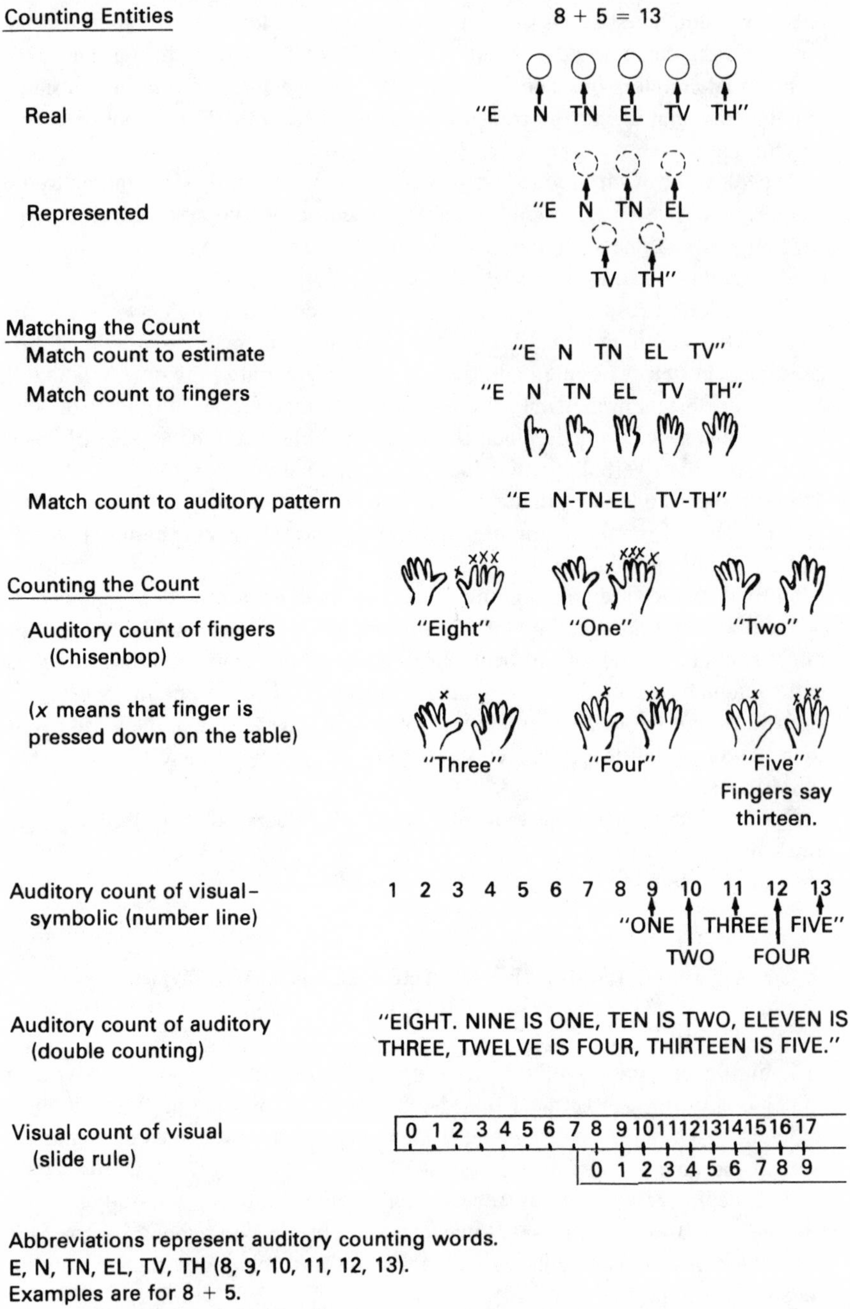

Counting Entities

8 + 5 = 13

Real — "E N TN EL TV TH"

Represented — "E N TN EL TV TH"

Matching the Count

Match count to estimate — "E N TN EL TV"

Match count to fingers — "E N TN EL TV TH"

Match count to auditory pattern — "E N-TN-EL TV-TH"

Counting the Count

Auditory count of fingers (Chisenbop) — "Eight" "One" "Two" "Three" "Four" "Five" Fingers say thirteen.

(*x* means that finger is pressed down on the table)

Auditory count of visual-symbolic (number line) — 1 2 3 4 5 6 7 8 9 10 11 12 13 "ONE TWO THREE FOUR FIVE"

Auditory count of auditory (double counting) — "EIGHT. NINE IS ONE, TEN IS TWO, ELEVEN IS THREE, TWELVE IS FOUR, THIRTEEN IS FIVE."

Visual count of visual (slide rule) — 0 1 2 3 4 5 6 7 8 9 10 11 12 13 14 15 16 17 / 0 1 2 3 4 5 6 7 8 9

Abbreviations represent auditory counting words.
E, N, TN, EL, TV, TH (8, 9, 10, 11, 12, 13).
Examples are for 8 + 5.

FIG. 6.1. Methods of keeping track in counting on.

subjects into three main categories: Counting Entities, Matching the Counting Words, and Counting the Counting Words (Fig. 6.1). In the Counting Entities category, a set of entities with a cardinality of the second addend is supplied or generated by the individual. The first addend word is produced, and then the set of entities is counted beginning with the count word following the first addend word. Counting on ends when the entities for the second addend have all been counted. The entities may be real objects (blocks, fingers, dots on paper) or they may be represented mentally. In the latter case, the entities may be arranged in a pattern to facilitate accurate and stable representation.

In the next two keeping-track categories, the counted-on words shift from being used to count other entities to themselves being the entities that are counted or matched. In the second category (Matching the Count) an auditory or a finger pattern of entities with the cardinality of the second addend is known or generated (e.g., for six: "da–da–da da–da–da" or one hand plus one finger). The first addend word is produced and then as each succeeding word is produced, one entity from the pattern is used up until the whole pattern has been completed. Thus the words and the pattern elements are matched, and a feedback loop occurs after the production of each word to check whether the entire pattern has been produced. When it has, word production ends. The difference between the use of fingers in this category and in the first is that here each finger is put up one by one as a word is produced until the required finger pattern is displayed. In the first category, six fingers would have been put out and all held out while those fingers were counted. The checking feedback loop with fingers may involve visual or kinesthetic feedback or both; some children clearly extend their fingers and look at them during this process, whereas others use very subtle finger movements. An early form of this Matching the Count method that we have termed "estimating" is used by some children. They count on and seem simply to stop producing words when they have produced "about enough" more words. These children do not employ a very definite pattern. This method may be an initial attempt to extend the subitizing method used for very small numbers to larger numbers. When a child finds that this estimated means is inaccurate, the child may then move to another keeping-track method.

The third category of keeping-track methods is more difficult; these methods were employed chiefly by adults. They employ some kind of double counting procedure in which the counted-on words are themselves counted (Fig. 6.1 and Fuson, 1981). The counted-on words may be auditory, visual–symbolic as in a number line, or represented on fingers as in the Chisenbop method. The counting process is auditory for all of these, with an additional fourth method achieved by the use of a visual count of a visual–symbolic counting sequence, such as in an addition slide rule. Some adults, for example, when asked to do an alphabet problem such as K + G would write out an alphabet and would then write out A B C D E F G above the letters L through R (the visual–visual method). Others would produce the A through G "counting words" as they pointed to the letters

L through R (the auditory–visual method). One brave student did the auditory–auditory pattern: L is A, M is B, N is C, . . . , R is G.

Table 6.1 reports the keeping-track methods used by the children in the addition solution procedures study described earlier. These data are from two conditions: the condition reported earlier in which the first addend card was facedown and the second addend card faceup, and another condition in which both addend cards were facedown. In the first of these conditions, second addend squares were available for counting but in the second condition they were not. When the squares for the second addend were available, about a third of the sample used the squares, but two-thirds of the children used some other method for at least one problem. The use of fingers was the second most popular method. Five children used an auditory–fingers double counting procedure. About half of the children in this sample had learned Chisenbop the preceding year, which may have encouraged the use of this method.

There does seem to be some relationship between counting-on performance on the first addend and keeping-track methods. All five children who used counting-all or transitional procedures for the first addend used the squares to

TABLE 6.1
Summary of Children's Keeping-Track Methods

Method	*Experimental Condition*	
	Second Addend Objects Visible	*Second Addend Objects Not Visible*
Counting Entities		
Real	8(28)[a]	not possible
Represented	2(5)	3(11)
Matching the Count		
Match count to estimate	1(1)	4(10)
Match count to fingers	7(24)	8(28)
Match count to auditory pattern	3(10)	3(8)
Counting the Count		
Auditory count of fingers (Chisenbop)	5(17)	5(18)
Auditory count of visual-symbolic (number line)	0(0)	0(0)
Auditory count of auditory (double counting)	1(1)	1(2)
Visual count of visual (slide rule)	0(0)	0(0)
Not clear, etc.	1(1)	3(8)

[a]*n(m): n* is the number of subjects out of 27 using the given method; some subjects use more than one method.

m is the number of problems (four of each kind were given) on which the method was used; responses involving procedures other than counting on are not included.

keep track when they could do so. When the squares were not available, three of these children used an estimate, one generated a visual representation for the second addend, and one child used fingers on some trials and a visual representation on others. The three children who used the squares when they were available, but who used counting on rather than the counting-all procedure when no squares were available, used fingers or double counting or solved the addition problem without counting by the use of a solution derived from a known addition fact. Thus, when forced into a condition in which they could not use the simplest keeping-track methods, the users of the more primitive first-addend procedures also used more primitive procedures for the second addend.

Counting real entities is clearly the simplest and the earliest keeping-track method; it involves a minimal change from the counting-all procedure. Counting represented entities, matching the count with fingers or auditory patterns, and auditorily counting the finger count of Chisenbop all seem to be used by somewhat more advanced children, and these methods are used in preference to counting real entities. Except for Chisenbop, all the Counting-the-Count methods seem to be relatively advanced (only one child used one of these methods) and perhaps less likely to be discovered spontaneously by a child. Adults did spontaneously use these methods, however.

Researching keeping-track methods is difficult because much of the execution of these methods is at least partially covert. Furthermore, individuals seem to differ in their preferred modes of imagery (visual, auditory, kinesthetic, body movement) and in fact use different combinations of these modes. Steffe, Thompson, and Richards (this volume) have focused on these differences in what is counted, or in what is the unit in the counting act, and have specified the following types: perceptual, figural, motor, verbal, and abstract unit items. These different types of unit items may be involved in what is counted or in what is matched in the keeping-track methods.

All the data we have reported on children's keeping-track methods concerns their spontaneous use of such methods. Studies using instructed conditions and explicit training of methods also need to be done for us to understand not only what methods children do use but also what methods they are capable of using if taught to do so.

COORDINATING THE FIRST AND SECOND ADDENDS IN COUNTING ON

Correct counting on requires that the procedures for each addend be correctly coordinated. In a study in which children were asked to count up a given number of steps from a certain number, observations of the spontaneous use of the fingers to keep track of the second number revealed a particular error in keeping track that was displayed on at least three trials by about a third of the sample (Fuson,

1981). This error was to put out the first keeping-track finger as the child produced the word that named the first (starting) number. Thus, in the problem "Start with 8 and count on 5 more," the child would begin by saying "eight" and would extend the first finger when "eight" was said. Successive fingers then were extended with each additional number word produced, up to a goal of five extended fingers. This always resulted in an answer that was one number too small. Several adults in a study in which the alphabet was used to stimulate counting procedures demonstrated a similar difficulty. When counting on for M + G, they reported that they had generated a finger representation for G (i.e., they had devised an appropriate keeping-track method), and they knew they should just say M and then say more letters (N, O, P, etc.) while using up their finger representation for G. However, they didn't know whether they should put up the first finger when they said M or when they said N.

In both of these situations, subjects were using counting-word sequences to solve an addition problem without any object referents for the counting words being used. Thus, there was little support for the cardinal meanings of the addends. The occurrence of this error in connecting the counting-word sequences for the two addends seems to imply that it is important for counting on first to be learned in an object context where the arrangement of the objects facilitates the correct coordination of the counting-word sequences for each addend.

This need for objects for correct addend coordination is contrary to the need discussed earlier for the first addend. The behaviors transitional between counting all and counting on seemed to be focused on the sequence of counting words and to be somewhat withdrawn from the object context. Perhaps a focus on the counting words facilitates the transition from counting all to counting on, but this transition may nevertheless have to occur within a context in which the cardinal referents for each addend are available and are not ambiguous.

CONCLUSION

Children in this country, in the Soviet Union, and in Sweden (Svenson, 1975; Svenson & Broquist, 1975) clearly use the counting-on procedure to solve addition problems. The time period during which children use this procedure is not so clear. Under certain conditions counting on can occur very early. In a study examining the development of the act of counting (Fuson & Mierkiewicz, 1980) a row of blocks was systematically increased in length and children were asked how many blocks there were each time blocks were added to the row. Three of 12 3-year-olds counted on for at least one trial when only one block was added. That is, they remembered the size of their previous count and said, for example, "Fourteen, fifteen. There are 15 blocks now." In that same task six of 12 5-year-olds, for at least 60% of their trials, counted on when one or two blocks were added and two more 5-year-olds counted on for at least 40% of their trials.

Five months later six of those 12 children counted on for at least 90% of their trials and four more for at least 60% of their trials. Groen and Resnick (1977) reported reaction-time measurements indicating that after 30 practice sessions, half of their sample of 4-year-olds counted on for sums less than 10.

The children in our studies used four different addition solution procedures: counting all, counting on, recall of facts, and solutions derived from known facts (e.g., $5 + 7 = 12$ because $6 + 6 = 12$). The use of these different procedures depends heavily on the particular addition problem given. The use of counting on replaces the use of counting all, and as the number of recalled facts grows, counting on may move from a principal means of sum finding to a "backup" means used when a certain addition fact cannot be recalled. Counting is also a useful procedure with larger addends. For a problem such as $13 + 6$, some children in our study counted on in the ordinary way ("13, 14, 15, . . . , 19"), whereas others separated the digit numeral from the teens and counted on from the larger number, saying, "six, seven, eight, nine, so it's 19." In doing sums involving two-digit numbers (e.g., $27 + 36$), children initially count on by ones (i.e., count on 27 words past 36) but some eventually devise or learn to count on by tens and by ones (e.g., 36, 46, 56, 57, 58, 59, 60, 61, 62, 63). Children thus may use counting on for a considerable portion of the first 3 years of elementary school and occasionally after that.

Object presence, object arrangement, size of addends, and amount of practice are variables that can affect counting-on performance. These variables will have to be carefully controlled in future research. As we come to understand more fully how the knowledge and skill components that underlie generalized and accurate counting on are acquired, we can begin to turn towards instructional issues. That is, we can then move from assessing the extent to which children already *do* understand and use counting on to the extent to which they *can* understand and use this procedure. Attention will also need to be turned towards examining possible negative effects of counting on (e.g., continued reliance on this procedure rather than an attempt to commit facts to memory).

Research on counting on can also address some issues of more general importance in early mathematics learning. Children probably spend a great deal of their early mathematics learning shifting back and forth among various meanings for number words and number symbols (Fuson & Hall, in press). This constant shifting of meanings may be an unrecognized source of learning problems. For adults, these meanings are all so well linked that the difficulty of such shifting for young children may be underestimated. Research into the shifting of meanings within the first addend word in counting on may begin to provide some insight into this general problem. Likewise, the study of the parts of counting on involving the use of the sequence of number words removed from an object context (such as in transitions from counting all to counting on or in the last two keeping-track procedures) may help to provide findings concerning such uses of the counting-word sequence generalizable to other mathematical procedures.

Both of these more general aspects of counting on may have important parallels in counting solution procedures used in subtraction, multiplication, and division operations on integers and on rational numbers.

ACKNOWLEDGMENTS

This chapter is a shortened version of a much longer paper (Fuson, 1981), which contains details of the design and results of the studies summarized here and a longer discussion of the procedures and results of Davydov and Andronov. The longer version can be obtained from the author at the School of Education, Northwestern University, Evanston, Il. 60201. I would like to acknowledge the debt my work in this area owes to periodic interactions with the work and thinking of Leslie Steffe. His insightful analyses of the solution processes of individual children and the creative tasks he has devised for examining the thinking processes of the child have contributed enormously to my own thinking and empirical work. Thanks also to Max Bell, James W. Hall, Eric Hamilton, Marsha Landau, Richard Lesh, Paula Olszewski, John Richards, Don Saari, Walter Secada, and the editors for providing helpful commentary on the several earlier versions of this chapter, and special thanks to Tom Carpenter for the work he did on the difficult task of shortening it. Walter Secada collaborated very effectively with me on the addition solution procedure study and is responsible for the data gathering and much of the data analysis in that study. Special thanks to Betty Weeks and to Leona Barth for arranging and coordinating our data-gathering efforts and to the teachers at the National College of Education Demonstration School and the Walt Disney Magnet School for their very considerable help. This material is based on work supported by the National Science Foundation and the National Institute of Education under Grant No. SED 78-22048. Any opinions, findings, and conclusions or recommendations expressed in this publication are those of the author and do not necessarily reflect the views of the National Science Foundation or the National Institute of Education.

REFERENCES

Davydov, V. V., & Andronov, V. P. *Psychological conditions of the origination of ideal actions.* (Project Paper 81-2) Wisconsin Research and Development Center for Individualized Schooling, The University of Wisconsin, 1981.

Fuson, K. C. *An analysis of the count-on solution procedure in addition.* Northwestern University, Evanston, Ill. Unpublished paper, 1981.

Fuson, K. C., & Hall, J. W. The acquisition of early number word meanings. To appear in H. Ginsburg, (Ed.), *Children's mathematical thinking.* N.Y.: Academic Press, in press.

Fuson, K. C., & Mierkiewicz, D. *A detailed analysis of the act of counting.* Paper presented at the annual meeting of the American Educational Research Association, Boston, April 1980.

Fuson, K. C., Richards, J., & Briars, D. J. The acquisition and elaboration of the number word sequence. In C. Brainerd (Ed.), *Children's Logical and Mathematical Cognition* (*Progress in Cognitive Development,* Vol. 1). Springer-Verlag, in press.

Gelman, R., & Gallistel, C. R. *The child's understanding of number.* Cambridge, Mass.: Harvard University Press, 1978.

Groen, C. J., & Resnick, L. B. Can preschool children invent addition algorithms? *Journal of Educational Psychology,* 1977, *69*(6), 645–652.

Schaeffer, B., Eggleston, V. H., & Scott, J. L. Number development in young children. *Cognitive Psychology,* 1974, *6,* 357–379.

Secada, W., & Fuson, K. C. *Addition counting solution procedures: Preliminary empirical findings.* In preparation.

Steffe, L. P., Spikes, W. C., & Hirstein, J. J. *Summary of quantitative comparisons and class inclusion as readiness variables for learning first-grade arithmetical content.* Working Paper No. 1, University of Georgia, Center for Research in the Learning and Teaching of Mathematics, 1976.

Svenson, O. Analysis of time required by children for simple additions. *Acta Psychologica,* 1975, *39,* 289–302.

Svenson, O., & Broquist, S. Strategies for solving simple addition problems: A comparison of normal and subnormal children. *Scandanavian Journal of Psychology,* 1975, *16,* 143–151.

7 Children's Counting in Arithmetical Problem Solving

Leslie P. Steffe
University of Georgia

Patrick W. Thompson
San Diego State University

John Richards
Massachusetts Institute of Technology

In this chapter, we examine the types of counting initially found in young children and describe how these counting types are related to children's solutions of addition and subtraction problems. We: (1) describe the first three counting types used by young children, all of which are prenumerical; (2) describe children's solutions to arithmetical tasks that involve these counting types; and (3) argue that the counting type a child can use is directly correlated with the child's method of solving arithmetical tasks.

We distinguish between different types of children's counting based on what it is that the child seems to generate and be aware of while counting. We call the objects the child creates *unit items*. This provides us with a working definition of counting:

Counting is the production of a number word sequence,
such that each number word is accompanied
by the production of a unit item.

There are five types of unit items—perceptual, figural, motor, verbal, and abstract.[1] An understanding of the type of unit children create in counting, their

---

[1]The classification of unit types is to some extent based on the analysis of numerical concepts carried out within our project by Ernst von Glasersfeld. Our analysis of unit types would not have been possible without his insightful analyses (von Glasersfeld, 1981).

counting type, is essential for understanding their construction of solutions to arithmetical tasks. Following is a description of the first three counting types.

Counting Types

Counters With Perceptual Unit Items. Whenever the child's countable items can be directly perceived, we speak of counting with perceptual unit items. We call a child a counter with perceptual unit items as long as that child requires the perceptual component and is unable to count unless a collection of perceptual items is actually available. Counters with perceptual unit items can count spatial arrangements (arrays as well as rows) of various items, and piles of items arranged in no particular manner. They also can count successive sounds like the ringing of a church bell as well as successive homogeneous physical actions like the to-and-fro swings of a pendulum. The crucial feature is that the perceptual items be actually constructible through some sensory mode.

Counters With Figural Unit Items. The first manifestation of independence from immediate perception in counting occurs when a counting episode involves items that are not within the child's range of perception or action. In this case we speak of counting with figural unit items. If a child's countable items are limited to perceptual and figural units, we call that child a counter with figural unit items. This is demonstrated in the following protocol. In this task from Steffe, Hirstein, and Spikes (1976)[2] three of the discs in a row of eight were screened. We inferred that Glen, a 6-year-old, was counting with figural unit items.

Interviewer: I'll let you feel that one right there (the first disc covered). There are a whole bunch of them on the board. That's the first one. You can't touch the others! Now, there are some counters under there. Now, get your hand off! (as Glen tried to touch the second covered disc). This is the first one (pointing to the first covered disc), this is the fifth one (pointing to the fifth in the row). Which one is this? (pointing to the sixth).

Glen: Sixth.

Interviewer: That's right. This is the fifth one (again pointing to the fifth). How many are there in all? (gesturing over the covered and visible discs).

Glen: Attempts to touch the covered discs again, but is stopped by the interviewer. He then looks at the cover, successively fixing his gaze on specific locations in a linear pattern across the cloth in rhythm with subvocally uttering "one, . . . , six" and slightly nodding his head (counting with figural unit items). He then finishes his counting activity, counting the visible checkers "seven, . . . , eleven. Eleven."

---

[2]A similar task has been used independently by the Russian researchers Davydov and Andronov (1981) in studying the internalization of counting actions.

Glen, having long developed the concept of object permanence, knew that some of the discs were screened, so he was able to construct figurative representations of the screened items. These representations may have been quite incomplete in that they probably did not replicate all the features of the discs like color, shape, or texture. Partial representatives could serve as items to be counted, provided the feature of discreteness was maintained. Glen's counting actions, which involved the cloth, were made possible by his intention to count the screened discs. We classified Glen as a counter with figural unit items on the basis of his solution to a variety of tasks. His most sophisticated counting activity is exemplified in the foregoing protocol.

Counters With Motor Unit Items. When a child's countable items are motor actions, we speak of counting with motor unit items. Whenever a child's countable items are limited to perceptual, figural, and motor units, we call that child a counter with motor unit items.

A good example of counting with motor unit items is provided in the following protocol from an interview with Brenda, a 6-year-old who was classified as a counter with motor unit items. She was presented a posterboard with 12 checkers glued to it, seven of which were covered by a cloth. The seven covered checkers were to Brenda's left and the five visible checkers were to her right and formed a circle.

Interviewer: There are seven checkers under this cover. How many checkers are there in all on the card?

Brenda: Begins by subvocally uttering "one, . . . , seven," while pointing her pencil in the air over the cover each time indicating a different location, but clearly not referring to a specific place on the cloth. Pauses, then subvocally utters "eight, . . ., twelve" while pointing her pencil at each of the visible checkers. "Twelve."

As Brenda pointed over the cover with her pencil, we believe she intended to count the screened checkers, because she pointed to distinct places on the cover. We also consider her to have formed some figural representation of the discreteness of those checkers. What she counted, in our opinion, were the motor acts of pointing the pencil.

LEVELS OF PROBLEM SOLVING

We have classified 34 mid-year first-graders according to the type of counting of which they were capable. Thirteen were classified in the first three types that are discussed in this chapter. In this section, we describe the problem-solving behavior of these children.

In order to consider that children are engaged in true problem solving, we require that they act as if they set a primary goal and subsequently set and satisfy subgoals in the satisfaction of the primary goal. In other words they have to plan and anticipate the effects of their actions. Some children may be limited to a script-based understanding of both the problems and the arithmetical operations. Schank and Abelson (1977) developed the notion of script in an attempt to characterize ways in which humans deal routinely with their worlds. A script, in their usage, is a number of causal chains of conceptual (language-free) events bound more or less loosely into a conceptual unit. There are going-to-work scripts (e.g., get into car; insert key; start engine; check toll-token holder), grocery-store scripts, and so on. Scripts are necessarily idiosyncratic to the script holder, and are built to handle commonly experienced situations. One understands by way of a script by imposing onto a situation the knowledge held in the script. Being limited to script-based understanding means that one ascribes no meaning to the situation beyond that intrinsic to the script itself. But even when children are not limited to script-based understanding, scripts may be a factor in our explanation of their behavior in problem-solving situations.

Counters with Perceptual Unit Items

Five children were identified as counters with perceptual unit items. These five children's solution methods for addition and subtraction problems were different in two ways from those of children who were capable of more advanced forms of counting. These differences are: (1) reliance on perceptual referents for quantitative words; and (2) limitation to a script-based understanding of both the problems and arithmetical operations.

The first characteristic does not surprise us. These children needed perceivable objects to count. The perceptual referents were either some objects (cars, marbles, buttons, etc.) or standard finger patterns. The second characteristic is more revealing of the meanings that these children gave to addition and subtraction and the understanding they had of the stories. The addition script that they seemed to be calling upon was:

Make an addend corresponding to the first number word heard.
Make another addend.
Count to find "how many."

The children's subtraction script was of the same nature as that for addition.

Make the minuend.
Take something away.
Count what's left.

In these descriptions, we use "addend" and "minuend" only for lack of better terms. We do not mean that these children constructed numerical meaning for number words. The primary function of the scripts for these children was not so much to give meaning to the problems per se, but to give meaning to the situation of an adult reading stories about how many to them. When the children saw themselves in such a situation, they formed an expectation about what they were going to be doing; they became ready to activate a script they had for how many. In order to understand a problem that they saw as one of addition or subtraction, they had to actually carry out the actions depicted in the story, filling in the meaning of the number words in the context of their script by taking as an addend either: (1) a given pile of perceivable objects; (2) a set of perceivable objects created through counting; or (3) simultaneous extension of fingers. Although the scripts, if applied correctly, could lead to correct solutions to addition and subtraction problems, the perception-bound reasoning frequently led to errors.

The following protocol indicates both the nature of the script that one child called upon and the apparent randomness of his reasoning. Chuck seemed to be satisfied that he had created a referent for the number words in the stories merely by seeing more relevant objects.

Interviewer: Bill has three marbles. Tom gives him five more. How many marbles does Bill have now?
Chuck: Would you mind saying that again?
Interviewer: (Repeats problem.) Do you want to use the marbles to help you work the problem?
Chuck: Yeah. (Takes ten marbles from box; looks at the interviewer.)
Interviewer: Bill has three marbles.

Chuck: Okay, three. (Removes 3 marbles from pile, one at a time.)
Interviewer: Tom gives him five more.
Chuck: (Moves all the remaining marbles to the 3 he had set aside.) One, two, three, . . ., ten. (Counts the marbles in the pile.)

After Chuck had satisfied his goal of "make three," he satisfied his next goal by looking at the remaining marbles; thus, he believed he had isolated the marbles corresponding to make another addend. If Chuck did indeed rely on the perception of an isolated group of objects as satisfying a "make-another-addend" goal, then counting should have been largely irrelevant for him in terms of goal satisfaction. This indeed seems to have been the case, as seen in the following protocol:

Interviewer: There are eight buttons in a bag. Jane takes two buttons out of the bag to sew on a dress. How many buttons are left in the bag?
Chuck: (Takes all ten buttons from the objects box.)

Interviewer: There are eight buttons in a bag.
Chuck: Uh–huh (yes).
Interviewer: How many buttons down there?
Chuck: Eight.
Interviewer: Count them.
Chuck: One, two, three, . . . , ten (pointing at each button in synchrony with verbal count).
Interviewer: How are you going to work the problem?
Chuck: You take five away. One, two, . . . , five (while removing one button in synchrony with each utterance; pause).
Interviewer: And?
Chuck: One, two, . . . , five (pointing at each of remaining buttons).

The most striking aspect of Chuck's protocol is that he seemed not to experience conflict when, after saying that there were eight buttons in his pile, he counted to ten.

These children's limitation to script-based understanding of the problems became most apparent when we examined their solutions to missing-addend problems. The following protocol was typical of the solution procedure for each of the counters with perceptual units.

Interviewer: (Places 10 blocks on the table.) Susan had five blocks. Her mother gave her some more blocks, and now she has eight. How many blocks did her mother give her?
Beverly: (Upon hearing "five" she counted out five blocks, separating them from the pile; re-counted the five blocks, and continued counting the blocks remaining in the original pile.) Ten.

The children's performances on the missing-addend problems seemed to be based on a generalization of the addition and subtraction scripts.

Make one pile.
Make another pile.
Count to find how many.

In missing-addend problems, the means of satisfying these goals varied between and sometimes within children. In the foregoing protocol, Beverly seemed to satisfy "make another pile" merely by noticing the remaining blocks. Others actually attempted to make another pile, but failed, either because of too few objects or too few fingers, and stopped in confusion.

It seems clear to us that these children were not engaging in problem solving, at least as problem solving is conventionally thought of. They did not attempt to satisfy a primary goal "how many" by setting and satisfying subgoals. Rather, they satisfied the goal "how many" because "count to find how many" was the last thing to do according to their script. This conclusion is based on their lack of differentiation between regular addition and missing-addend problems and the random way in which they solved both.

Counters with Figural Units Items

The first manifestation of independence from immediate perception in counting involves counting with figural units. Three children in our study were classified as counters with figural unit items. When solving problems with objects available, they seemed to be operating according to the same situational script as counters with perceptual units. They were unlike the more primitive counters, however, when solving problems without objects available; at times they seemed to engage in true problem solving.

We inferred that on at least one problem each of the counters with figural units had created a set of related goals whose satisfaction, in their view, constituted a solution to the problem. Their solutions also differed from those of the counters with perceptual unit items in that they could create referents in the absence of perceivable objects. To solve verbal problems, however, these children still relied on finger patterns to serve as records of their constructive acts. But it was the constructions that seemed to be of primary importance. The finger patterns seemed necessary only as records; these children were unsuccessful in coordinating successive steps of their solution whenever they failed to generate finger patterns. The following protocol illustrates this point.

Interviewer: Kevin has four crayons. Jerry gives him three more crayons. How many crayons does Kevin have now?

Greg: Ten.

Interviewer: You don't have to do this in your head. You can use your fingers. How many does he have now?

Greg: Six.

Interviewer: How did you do it?

Greg: I just think it.

Interviewer: Did you add or take away?

Greg: I add.

Interviewer: Write down what you added.

Greg: (Writes "4 + 3".)

Interviewer: What did you get when you added four plus three?

Greg: It goes eight.

Interviewer: Check it out. Use your fingers. Think what four plus four is.

Greg: Four plus four is . . . three plus three is six. So four plus four is seven.

Interviewer: Four plus four is seven. Then what is four plus three?

Greg: Eight.

Interviewer: So look, four (points to his sentence)—you put three things onto four.

Greg: (Simultaneously extends all ten fingers; folds two fingers on one hand, one on the other.) One, two, three, . . . , seven (in synchrony with wiggling each extended finger).

It was not until Greg displayed four and three fingers, that he was able to solve the problem, in our view if not his. After he realized he could use his fingers, he solved another problem:

Interviewer: Mike has three cats. His mother gives him some more. He now has seven. How many did his mother give him?

Greg: (Writes "7 + 3 = ______"; simultaneously extends five fingers on left hand; pauses; sequentially extends two fingers, on right hand; pauses; continues, sequentially extending three more fingers on right hand in synchrony with subvocally uttering "8 - 9 - 10".). Ten!

Greg's counting in the tasks used to classify him as a counter with figural units never included sequential or simultaneous finger extensions. This limitation essentially prohibited him from using a counting technique in problem solution until the interviewer suggested that he use his fingers. After creating finger patterns in this solution, he appeared to realize that his fingers could be used to solve other tasks. He solved the next addition problem (3 + 5) immediately. After writing "3 + 5 = ______," he simultaneously extended three fingers on one hand and then five on the other. He then counted in a way used by none of the four counters with perceptual unit items. He separated the two finger patterns. After extending the five fingers, he immediately counted the three fingers, "six, seven, eight—eight." Although the fingers served as perceptual items to count, his creative use of counting indicates that he was not a slave to his extended fingers.

The five fingers served the role of a perceptual pattern. This allowed him to appear precocious in counting on; the five fingers serving as a starting place for counting. His linguistic system for uttering number words then propelled him forward as he counted the remainder of his fingers. However, it would be an understatement to say that he did not go beyond counting with perceptual unit items as he uttered "six, seven, eight." His protocol of the last problem shows he finally created motor unit items. The fact the motor unit items were not spontaneously produced, but were a result of directed interference by the interviewer, distinguishes Greg's behavior from that of children who would be classified as counters with motor unit items.

The second protocol indicates that Greg was actually solving a problem, as opposed to following a situational script. The goal structure he created, however, was quite different from what the interviewer intended. Nevertheless, writing an addition sentence (7 + 3 = ______) signifies that he did create a goal structure before counting. His most general goal was to construct a referent for the number of Mike's cats. To achieve this goal, he created and satisfied a subgoal of representing seven by extending seven fingers. This subgoal was related to his second subgoal (representing three) by constructive counting actions.

While satisfying his first subgoal, he deferred his general goal and his second subgoal. Upon achieving the first subgoal, he held it in memory, his fingers serving as a record. He then revisited his second subgoal but was not required to count because three was part of his subitizing system.[3] Because both subgoals had been satisfied, he was able to again consider his general goal. Having three things to count, he was capable of continuing counting as he would if he had extended three fingers and started counting all over again from one. It was in this sense that the two subgoals were related through counting actions. It is important to note that he did not simultaneously satisfy his second subgoal and his general goal. His strategy was probably of a "hill climbing" nature (Simon & Newell, 1971)—pauses between episodes indicated to us that he was asking himself "What do I do now?" as each subgoal was satisfied.

There is evidence from other interviews that Greg constructed goals prior to problem solution. For example, in an interview using a missing-addend problem without objects (5 + ______ = 8), Greg wrote the sentence 5 + 8 = ______ and attempted to solve this sentence. However, in all interviews where objects were available, he appeared not to create general goals. The presence of objects did not appear to be the only source of confusion; he solved every problem without objects as if it were an addition problem whether it was addition or not. These distortions do not mean he did not create a goal structure in some cases, they mean only that an additive goal structure was the only one Greg was capable of when objects were not available. When objects were available, he was more sensitive to the problem statements, but still gave no evidence of a goal structure for the missing-addend and subtraction problems.

These three counters with figural units were not consistently good problem solvers. The protocols we have cited are instances of what they were capable of once the interviewer probed deeply enough. On the word problems with objects available, they often performed as if they were limited to situational scripts, just like the counters with perceptual unit items. Unlike the counters with perceptual unit items, they could tentatively extend counting to counting with motor unit items under the right circumstances. When they did, they seemed to make a problem of the situation. But these children had not fully developed their schema for counting with motor unit items and could not operate with motor units that they could not record on their fingers like gestures.

Counters with Motor Unit Items

A counter with motor unit items is aware of the motor act as a discrete experience. This awareness allows the counter to execute counting actions independently of perceptual items, even though that may be what he intends to count. An

[3]Subitizing is the perceptual recognition and discrimination of patterns of small lots consisting of up to five or six items (Beckwith & Restle, 1966; Brownell, 1928; Gelman & Gallistel, 1978; Kaufman, Lord, Reese, & Volkman, 1949; Klahr & Wallace, 1973; Schaeffer, Eggleston, & Scott, 1974; von Glasersfeld, 1980).

essential feature of counting with motor unit items is that the motor acts must be generated with the intention of counting in the absence of directly perceivable objects.

Two outstanding characteristics separated the five children classified as counters with motor unit items from the three children discussed in the preceding section. Each of them created a goal structure that was freed from perceptual material prior to solving at least one arithmetic problem. Moreover, all five children spontaneously carried out motor actions, usually finger extensions. These two characteristics are in consonance with the children's capabilities: (1) to anticipate the construction of an item that can be counted that is independent of perceptual material; and (2) to search actively for a variety of motor actions. These five children were almost universally limited to motor action in counting when solving arithmetical problems. Their performance never indicated knowledge of basic addition or subtraction facts, counting on, or advanced strategies for finding sums or differences from known facts. When counting, they always counted in a forward direction, either starting with one or with a perceptual constant such as five fingers on one hand.

For a counter with motor unit items, the nature of counting essentially prevents the formulation of plans in solving arithmetical problems. In the case of addition problems, however, the children created a goal structure prior to initiating their solution. It seems to us the children had somehow constructed a script for addition that was freed from perceptions. The script seemed to be as follows, where "addend" is again used prenumerically:

Construct addend.
Construct addend.
Find how many.

We have considered the influence of available objects used by the children during their solution of subtraction problems. Although the children had the potential to create goals through script-based understanding, they seemed to prefer to create an idiosyncratic perceptual referent for the number word for the minuend in the problem statement. The following two protocols of Jamie illustrate this:

Interviewer: There are eight buttons in a bag (Jamie begins taking buttons from the box).
Jamie: (Takes buttons from box.) There aren't eight buttons here.
Interviewer: Count the buttons.
Jamie: (Puts all buttons in left hand; removes them one by one in synchrony with uttering "1-2- . . . -10.")
Interviewer: There are eight buttons in a bag. Jane takes two buttons out of the bag to sew on a dress (Jamie reaches for buttons).
Jamie: (Takes two buttons from the pile and places them on cloth.)

Interviewer: How many buttons are left in the bag?
Jamie: (Picks up two from those remaining in pile.) Two. (Continues taking buttons one by one.) Three, four, . . . , eight. Eight! She had eight left.

Interviewer: There are eight marbles in a bag. Jane takes two marbles out to play with. How many marbles are left in the bag?
Jamie: How many were in the bag?
Interviewer: Eight (no marbles were available).
Jamie: Jane took out two.
Interviewer: (With Jamie.) How many were left in the bag?
Jamie: (Spreads fingers of left hand; counts fingers with index finger of right hand.) One, two, . . . , five. If you took away two, then I'm gonna have to do it my way! (Counts fingers again, subvocally uttering (1–2–. . .– 5; sequentially extends fingers on right hand.) Six, seven, eight . . . Took away two (folds two fingers of left hand). Okay, those two are down. Now let's see (counts remaining fingers). One, two, . . . , six.
Interviewer: Okay.
Jamie: Okay? Six.

In view of her behavior in the first protocol, we infer that Jamie did create a perceptual referent corresponding to the statement, "There are eight buttons in a bag." That she realized that she had 10 buttons is corroborated by her statement "There aren't eight buttons here," and by the fact that she counted the 10 buttons. But her perception of the buttons seemed to overwhelm any intention she may have had to construct a pile of eight. She apparently never counted out a referent for the statement "There are eight buttons in a bag." Her counting actions were executed in response to a directive by the interviewer. Moreover, she experienced no apparent conflict between hearing "eight" in the problem statement and the results of her counting actions. This lack of conflict corroborates the claim that she created an idiosyncratic perceptual referent for the statement "There are eight . . . ," for had she intended to create any other referent, the results of her counting actions surely would have caused conflict. As it was, her goal appeared to be totally satisfied through visual perception of the buttons.

Although it would be a legitimate objection to say that the problem was not read to Jamie in its entirety prior to her solution, she solved a similar subtraction problem (7 − 5) in the same way when she was read the whole problem before she initiated a solution. The only variation in her solution was that she did not count the ten objects to start with. In the second protocol, she was forced to construct a referent for the statement of the minuend by counting. There were no readily available objects for use.

An important construct in describing the problem-solving behavior of children who could count with motor units is intuitive extension. Intuitive extension can be thought of as the continuation, after a pause, of an initial sequence of counting acts. The continuation is produced by counting with motor unit items. But this concept is more general. For example, in solving ''How much is six and two?'' a child sequentially extends fingers, uttering ''one, . . . , six,'' pauses, and then proceeds, uttering only ''seven, eight.'' Four of the children who could count with motor units were capable of intuitive extension. Maurice, however, was the only child who used intuitive extension in solving addition problems, and that was to justify a solution. Maurice's solution is described later. Charles, who was the most advanced of the counters with motor unit items, solved all four addition problems by creating a set of related goals through script-based understanding. However, in satisfying his general goal, he always performed his counting actions anew, starting with ''one,'' and never realized that he could use intuitive extension.

Missing-Addend Problems. At the first two levels of counting, children are unable to solve missing-addend problems as they are conventionally thought of. Our analysis suggests that the solution of a missing-addend problem would require intuitive extensions. We believe that children who are incapable of intuitive extension are also incapable of viewing a sum as the extension of one set of counting actions beyond another prior to initiating a solution. It seems to us essential that a child be able to view a sum in this way to plan a solution to a missing-addend problem, as the plan involves holding the goal corresponding to the sum in mind while constructing the extension beyond the known addend. Without intuitive extension, children would most likely interpret a missing-addend problem using their script for addition.

The following protocol of Brenda, a child who could count with motor unit items but showed no evidence of intuitive extension, illustrates our claim of the necessity of intuitive extension to solve missing-addend problems:

Interviewer: Tom had five comic books. He got some more for his birthday. Now he has eight comic books. How many more did he get for his birthday?
Brenda: He got eight for his birthday.
Interviewer: No. He had five, and now he has eight. How many more did he get? He didn't get them all for his birthday.
Brenda: Oh.
Interviewer: He already had five.
Brenda: He got eight for his birthday.
Interviewer: Did he? See, he only has eight after he is done. He had five to start with and then he got some more and now he has eight. How many more did he get?
Brenda: (Makes five marks with a pencil, then eight, subvocally counting. She then counts them all and writes $5 + 8 = 13$.)

Brenda's method of solution reflected her lack of intuitive extension. When objects were available for use, Brenda still could not create a goal structure and plan a solution in consonance with the problem structure.

The second issue we address is whether a child with intuitive extension is sufficiently powerful to create a goal structure and plan a valid solution for a missing-addend problem. One of the children whose performance allowed us to infer intuitive extension was Maurice:

Interviewer: Tom had five comic books.
Maurice: (Simultaneously extends five fingers on his left hand before the interviewer can go on.)
Interviewer: He got some more for his birthday. Now he has eight comic books.
Maurice: (Sequentially extends three fingers on his right hand.)
Interviewer: How many did he get for his birthday?
Maurice: Three-eight (Writes "8".)
Interviewer: (Repeats the question.)
Maurice: Three.

Our interpretation is that Maurice never created a goal structure or planned a solution prior to solving the problem. He created a referent for the statement of the known addend before hearing the second and third statements in the problem. He constructed the referent for the second addend by rapidly counting forward "six, seven, eight"; he stopped at eight because this goal he was given. His advance over Brenda is evident in that he was able to create and satisfy related goals corresponding to the second and third statements. His goal associated with "eight" served as a criterion toward which Maurice worked to satisfy the goal associated with "some more." He was able to separate his counting actions, maintaining his record of the known addend while continuing to count. We do not claim that five was included in eight by Maurice, but rather that he viewed his actions corresponding to five as part of the actions necessary to construct eight, but ones that were separate from the latter actions. Moreover, we do not claim that he was necessarily aware of an increase in the sense of adding one more each time he executed a counting act. We do believe, however, that he was aware of going beyond counting actions already executed.

Although intuitive extension allowed Maurice to create a goal structure in consonance with the problem structure, he did not view eight a priori as the extension of one set of counting actions beyond another. His solution plan, whatever it was, seemed to develop during problem solution. If Maurice had heard the entire problem before he began, he might have solved it without prompts from the interviewer and with a clear plan for solution.

The following interview with Bryan, another child capable of intuitive extension, is the most powerful solution of a missing-addend problem we observed among the counters with motor units:

Interviewer: Mike had three cats. His mother gave him some more. He now has seven. How many did his mother give him?
Bryan: Okay. He has three (simultaneously extending three fingers).

Interviewer: And he winds up with seven.

Bryan: (Engages in a confused set of finger extensions; ends with three fingers extended.)

Interviewer: And he winds up with seven.

Bryan: (Counts the extended three fingers, then sequentially extends the fourth and fifth fingers of the same hand; simultaneously extends four fingers on his other hand and continues counting two of them.) And he (looks at his hands) gets four more!

When the interviewer read the problem, Bryan readily created and satisfied a goal corresponding to the known addend by simultaneously extending three fingers. After the interviewer first stated "And he winds up with seven," Bryan attempted to understand the problem by means of finger extensions. But he failed to create either a goal structure or a plan for solution. The interviewer's prompt was given while Bryan was actively attempting to create a goal structure. His statement allowed Bryan to organize his finger extensions, but not with reference to the original problem statement. The statement, provided during Bryan's constructive activity, allowed him to create a goal with regard to his immediate activity. The only carry-over from the original problem statement was the record of satisfying a goal corresponding to the statement of the known addend.

Counters with motor units showed advanced forms of problem construction, understanding, and solution. They also showed, however, that they were not far removed from their days as counters with perceptual unit items. When they found themselves in a standard subtraction situation, they seemed to fall back on well-learned scripts. In addition situations they were not limited to direct perceptual representation of the problems, but they still relied heavily on scripts developed in earlier days. In classroom situations they may have needed no more than this to get by. It is apparent to us that they were in fact capable of far more, even with the limited resources they had developed.

SUMMARY

In this chapter we have argued that our classification of children in terms of the types of items they can count—either perceptual, figural, or motor units—is correlated with specific kinds of solutions for arithmetical tasks. Counters with perceptual unit items, restricted to solving problems with objects available, were confined to a single script-based strategy with no flexibility and demonstrated little understanding of the problem context from the observer's perspective. Counters with figural unit items had a minimal ability to represent items not directly available. They showed a minimal understanding of some problem contexts and in these cases used appropriate script-based strategies. Counters with motor unit items were able to construct primitive strategies for solving problems.

This depended essentially on their ability to maintain a distinction in the counting activity while continuing to count. It is again essential to point out that these children were all prenumerical, yet one was still able to see a growing awareness of the use of counting, and a growing power of their construction of numerical operations.

ACKNOWLEDGMENTS

This paper is based on work supported by the National Science Foundation under Grant No. SED78-17365. Any opinions, findings, and conclusions or recommendations are those of the authors and do not necessarily reflect the views of the National Science Foundation.

REFERENCES

Beckwith, M., & Restle, F. Process of enumeration. *Psychological Review,* 1966, *73*(5), 437-444.

Brownell, W. A. *The development of children's number ideas in primary grades.* Chicago: University of Chicago Press, 1928. (Supplementary Educational Monograph)

Dayvdov, V. V., & Andronov, V. P. *Psychological condition of the origination of ideal actions.* (Project Paper 81-2). Wisconsin Research and Development Center for Individualized Schooling, The University of Wisconsin-Madison, 1981.

Gelman, R., & Gallistel, C. R. *The child's understanding of number.* Cambridge, Mass.: Harvard University Press, 1978.

Kaufman, E. L., Lord, M. W., Reese, T. W., & Volkman, J. The discrimination visual number. *American Journal of Psychology,* 1949, *62,* 498-525.

Klahr, D., & Wallace, J. G. The role of quantification operators in development of the conservation of quantity. *Cognitive Psychology,* 1973, *4,* 301-327.

Schaeffer, B., Eggleston, V. G., & Scott, J. L. Number development in young children. *Cognitive Psychology,* 1974, *6,* 357-379.

Schank, R., & Abelson, R. *Scripts, plans, goals, and understanding.* Hillsdale, N.J.: Lawrence Erlbaum Associates, 1977.

Simon, H. A., & Newell, A. Human problem solving: The state of the theory in 1970. *American Psychologist,* 1971, *26,* 145-159.

Steffe, L. P., Hirstein, J., & Spikes, C. *Quantitative comparisons and class inclusion as readiness variables for learning first-grade arithmetic content.* PMDC Report #9, University of Georgia, 1976.

von Glasersfeld, E. The conception and perception of number. In S. Wagner & W. Geeslin (Eds.), *Modeling mathematical cognitive development.* Columbus, Ohio: ERIC/SMEAC, 1980.

von Glasersfeld, E. An attentional model for the conceptual construction of units and number. *Journal for Research in Mathematics Education,* 1981, *12,* 83-94.

8 The Development of Addition and Subtraction Abilities Prior to Formal Schooling in Arithmetic

Prentice Starkey
Rochel Gelman
University of Pennsylvania

Young children with no formal schooling in arithmetic possess some understanding of addition and subtraction. In this chapter we summarize research on the young child's understanding of addition and subtraction. We then examine the ability of Piagetian theory to account for these research findings. Finally, we consider some of the consequences of assuming that the abilities to count and to take advantage of counting procedures represent a universal cognitive ability.

SOME EARLY COMPETENCIES

Solutions to "Number-Fact" Problems

Until recently, the majority of research on early addition and subtraction abilities dealt with the young child's ability to compute with natural numbers. The most commonly used problems were drawn from the set of basic addition and subtraction facts. Early researchers (Clapp, 1924; Knight & Behrens, 1928; Wheeler, 1939) generally used 6- and 7-year-olds and sought to establish the relative difficulty of various number combinations for children of this age. Theoretical interest centered on whether learning any particular arithmetic problem or type of problem would interfere with or facilitate learning any other particular problem. The focus was on the interaction between laws of learning and a portion of the structure of arithmetic.

The strongest conclusions that can be drawn from these early studies are that the difficulty of addition and subtraction problems increases with numerosity and that learning a particular solution does not interfere with learning a similar solution. There is also evidence that learning to solve addition and subtraction problems together rather than separately facilitates acquisition (Buckingham, 1927). More recent work, generally using concrete rather than written materials (Bjonerud, 1960; Dutton, 1963; McLaughlin, 1935; Richard, 1964; Williams, 1965), has also demonstrated that 4- and 5-year-olds can solve some of the simpler number-fact problems, especially those involving only small numbers.

Much of the early and more recent number-fact research has been theoretically narrow. Researchers have been insensitive to the underlying processes employed by children in solving addition and subtraction problems or have even misrepresented these processes. As a consequence, the studies have often focused on whether the solutions obtained by the children were correct or not, but not on the underlying basis of solutions or the types of errors made. However, these shortcomings do not make number-fact research uninteresting; such research shows that young, unschooled children can, in an unspecified way, solve some simple addition and subtraction problems. The main difficulty is that these investigations leave us almost totally in the dark as to the basis of the child's solutions. Only recently has a different type of research shed light on children's solution procedures.

Counting and Counting Algorithms

That preschool children possess a counting scheme and thus are capable of more than rote counting has been discussed elsewhere (Gelman & Gallistel, 1978), so we do not do so here. However, it should be noted that the use of counting algorithms by children can reveal something about the child's understanding of counting beyond mere application of counting to solve addition and subtraction problems. For example, the child who can use counting algorithms successfully to solve addition and subtraction problems must be able to recognize and select different types of counting in order to solve particular types of problems. The selection of backward counting to solve a subtraction problem and the selection of forward counting to solve an addition problem exemplify this. Counting on from a particular cardinal value or beginning with a cardinal value other than *one* also reveals something about the child's understanding of the count sequence. The use of recursive counting is also an interesting counting ability revealed by the use of certain counting algorithms. Thus, the nature and use of particular counting algorithms by young children are at least as informative about counting per se as about the child's understanding of addition and subtraction.

Although much of the early research only investigated item difficulty, occasional early reports (McLaughlin, 1935; Mott, 1945; Woody, 1931) mentioned the use of counting algorithms by children to solve addition and subtraction problems. One of the most detailed early attempts actually to describe some counting algorithms and link them to particular types of errors made by young children is found in a study by Ilg and Ames (1951). They found that the use of counting algorithms was prevalent in 5-year-olds attempting to solve addition and subtraction problems involving arrays of concrete objects and that the counting algorithms changed and became more efficient between the ages of 5 and 8 years. Ilg and Ames report that, to solve addition problems, the 5-year-olds counted from the number one up to the cardinal value of one of the numbers to be added; then, they counted on from this cardinal value for a number of steps equal to the cardinal value of the other number to be added. Fingers were counted when objects were not available, and incorrect answers usually deviated from the correct answer by $\pm$ 1. In contrast, the 6-year-olds started with the cardinal value of the larger of the numbers to be added and counted on for a number of steps equal to the cardinal value of the smaller number to be added.

To solve subtraction problems, the 5-year-olds chiefly counted from the number one up to the cardinal value of the minuend and then counted backwards for a number of steps equal to the cardinal value of the subtrahend. Older children (6- and 7-year-olds) counted backwards from the minuend for a number of steps equal to the cardinal value of the subtrahend. At older age levels, counting algorithms apparently underwent further change, though whether and how schooling influenced the change is unclear. The Ilg and Ames report is interesting in that it describes some particular algorithms used by children; however, the study is difficult to evaluate because the method and results are detailed inadequately. A particular shortcoming is the failure to report the frequency of occurrence of algorithms supposedly typical of a particular age level. Also, children younger than 5 years of age apparently were not investigated.

In the last decade, studies on counting algorithms (Groen & Parkman, 1972; Groen & Poll, 1973; Svenson, 1975; Svenson & Broquist, 1975; Svenson, Hendenborg, & Lingman, 1976; Woods, Resnick, & Groen, 1975) have begun to appear in the literature. Groen and Parkman (1972) found that 6-year-olds apparently solve simple written addition problems by counting on from the larger of the numbers to be added, an identical finding to that of Ilg and Ames. Woods, Resnick, and Groen (1975) found that second-grade children solve simple written subtraction problems apparently either: (1) by counting backward from the minuend for a number of steps equal to the cardinal value of the subtrahend; or (2) by recursive counting whereby the child counts on from the subtrahend until a cardinal value equal to that of the minuend is reached, and, while counting on from the subtrahend to the minuend, the child also counts the number of counting steps or increments required to reach the minuend value. This recursive counting algorithm evidently was not observed by Ilg and Ames.

In a closely related study, Groen and Resnick (1977) taught 4½-year-olds to solve simple addition problems by using a counting algorithm that consisted of counting two groups of objects equal in numerosity to the cardinal values of the two numbers to be added. Next, the groups of objects were combined, and the numerosity of the combined group was established by counting. Across sessions, half of the children spontaneously employed a more efficient algorithm they had not been taught by the experimenter. This new algorithm consisted of counting on from the cardinal value of the larger of the numbers to be added. Thus, Groen and Resnick provide important evidence that some young (4½-year-old) children spontaneously employ an efficient algorithm heretofore not reported in children younger than 6 years of age.

For more information on the nature, development, and use of counting algorithms, the reader is referred to the chapters by Carpenter and Moser and Fuson in this volume.

The finding that preschool children are capable of employing counting algorithms integrates nicely with several other findings concerning early number concepts. For example, Ilg and Ames (1951) and Thyne (1954) have reported that children's addition and subtraction errors commonly deviate from the correct answer by ± 1, an error pattern suggesting the use of imperfectly executed counting (Gelman & Gallistel, 1978). Also, the principal finding of number-fact research—that problem difficulty increases as numerosity increases—is consistent with the finding that preschool children are more likely to commit counting errors as numerosity increases (Gelman & Gallistel, 1978). Another common finding—that addition is easier than subtraction for young children—is also consistent with the use of counting algorithms, as backward counting is relatively difficult and is used in some subtraction algorithms but not in addition algorithms.

A study we recently conducted also fits with the finding that young children use counting algorithms (Starkey & Gelman, in preparation). We conducted the study to determine whether young preschool children were capable of solving a variety of problems in which objects were added to or subtracted from arrays that were screened from the child's view. Of special interest was the issue of whether the younger subjects would use counting algorithms to solve problems of this type.

A total of 48 children (16 3-year-olds, 16 4-year-olds, and 16 5-year-olds) participated in the experiment. Each of the tasks began by having the child establish the number of pennies held in the experimenter's open hand. The child was asked, "How many pennies does this bunch have?" The experimenter then screened the array of pennies and: (1) placed another array in the hand holding the original array (stating, "Now I'm putting *n* pennies in my hand; how many pennies does this bunch have?"); or (2) removed a subset of the array and displayed it to the child (stating, "Now I'm taking *n* pennies out of my hand; how many pennies does this bunch have?"). The initial array and the addend or subtrahend array were never simultaneously visible; because the arrays were covered, the children could not directly count the pennies to find the answers.

We found that preschool children (especially 4- and 5-year-olds) persisted in the use of counting algorithms when the objects to be counted were screened from view. Some children used fingers to represent the screened object; others counted aloud, apparently either imagining objects or working directly off the sequence of number names. Also, some children did not count overtly, though they gave correct answers on large number problems (e.g., 14 + 1 = ______), problems that they probably would not have memorized previously. A third group of children did not count overtly, did not solve large number problems, but did solve some small number problems. We do not know whether this group of children solved the problems by using counting algorithms or by using previously memorized number facts. However, enough overt counting and solutions to large number problems were observed to indicate that some preschool children (including a few 3-year-olds) can use counting algorithms even when the arrays that are to be added to or subtracted from are screened from view. As can be seen in Table 8.1, the 3-year-olds were capable of solving the easiest of problems (i.e.,

TABLE 8.1
Mean Proportions of Correct Answers on Number-Fact Problems

Simple Addition						*Simple Subtraction*					
	Addend Numerosity						*Subtrahend Numerosity*				
Augend Numerosity	*1*	*2*	*3*	*4*	*5*	*Minuend Numerosity*	*1*	*2*	*3*	*4*	*5*
3-year-olds											
1	.87	.53	.40	.27	.13	2	.87				
2	.73	.27	.13	.07		3	.40	.53			
3	.47	.27	.07			4	.27	.53	.13		
4	.33	.07				5	.27	.47	.20	.33	
5	.20					6	.27	.07	.33	.33	.40
14	.13					15	.13				
24	.07					25	.00				
4-year-olds											
1	1.00	.81	.69	.75	.44	2	1.00				
2	1.00	.50	.19	.25		3	.75	.62			
3	1.00	.50	.19			4	.75	.38	.69		
4	.94	.50				5	.69	.44	.44	.44	
5	.88					6	.56	.19	.12	.31	.38
14	.31					15	.19				
24	.13					25	.13				
5-year-olds											
1	1.00	.88	.88	.88	.88	2	.94				
2	1.00	.88	.44	.56		3	.94	.94			
3	1.00	.56	.69			4	.94	.44	.62		
4	.94	.81				5	.94	.68	.62	.56	
5	1.00					6	.88	.56	.56	.44	.44
14	.62					15	.44				
24	.88					25	.62				

$1 + 1 =$ ______, $2 + 1 =$ ______, and $2 - 1 =$ ______). Gelman and Gallistel (1978) report spontaneous counting on the part of 3-year-olds in simple addition and subtraction problems with arrays that were visible to the children. Taken together, these findings suggest that even younger children should be tested on very simple addition and subtraction problems, and that very young children may use some counting algorithms.

Basic Effects of Addition and Subtraction on Numerosity

To characterize adequately the child's understanding of addition and subtraction, one wants evidence beyond the fact that the child is capable of solving problems by the use of counting algorithms or some other type of algorithm. One wants evidence as to whether the child reveals an implicit understanding of some of the basic definitions and properties of arithmetic. Addition and subtraction of natural numbers are usually initially defined for young children in terms of operations on arrays of objects. Addition involves the combining of two arrays, subtraction involves removing some objects from an array of given numerosity. Understanding addition and subtraction of natural numbers must be based on an implicit understanding of the corresponding operations on arrays. Two of the most basic principles of arithmetic operations are that addition increases numerosity and subtraction decreases numerosity. These basic principles are known at least to some extent by preschool children. Smedslund (1966a) demonstrated this by presenting screened arrays of equal numerosity to 5- and 6-year-old children. When one of the arrays was transformed by adding one object to it or by subtracting one object from it, the children were able to designate correctly which posttransformation array contained more objects. The same findings were obtained with 4- and 5-year-olds by Brush (1978), with 3-, 4-, and 5-year-olds by Cooper, Starkey, Blevins, Goth, and Leitner (1978), and comparable addition results were found by Beilin (1968) and Mehler and Bever (1967). Related findings by Gelman (1972a, 1972b, 1977), and by Cooper et al. (1978) indicate that 3-, 4-, and 5-year-olds can infer the occurrence of a screened addition or subtraction transformation by comparing the pre- and posttransformation numerosities of arrays. Thus, preschool children have an understanding of the directional effects on numerosity of adding or subtracting elements from an array of objects, and they can infer the occurrence of either operation by establishing directional effects on numerosity. The earliest age at which the child understands the effects of these operations and can infer their occurrence is still unknown.

Arithmetic Laws: Inversion and Compensation

Another aspect of understanding the principles underlying addition and subtraction of natural numbers involves the child's use of basic properties of arithmetic operations to solve problems. Due to the importance Piaget ascribed to inversion and compensation, a few studies have examined their development. However, the development of other properties such as commutativity and associativity have rarely been studied.

By inversion, we mean that adding a particular number of elements to an array can be negated by subtracting the same number of elements. This property is reflected in the inverse relation between addition and subtraction ($a + b - b = a - b + b = a$). By compensation we mean that if the initial numerical relation between two arrays is altered by adding (or subtracting) elements to one of the arrays, the original relation may be reinstated by adding (or subtracting) elements to the other array. The corresponding properties for numerical operations are called the addition property of equality or the addition property of inequality depending on the relation involved (i.e., if $a = b$, then $a + c = b + c$ or if $a < b$ then $a + c < b + c$).

A distinction needs to be drawn between a child's ability to solve problems that might be solved using a particular property as opposed to the child's explicit knowledge of the property. This distinction can be illustrated by an example. A child with accurate counting algorithms should be able to solve problems such as $5 + 2 - 2$ by sequential calculations ($5 + 2 = 7$, $7 - 2 = 5$), whether or not explicit knowledge of the inverse operation was available to the child. Explicit knowledge of inversion would only be involved if a child recognized that the answer must be 5 because the same number was added and subtracted.

A relatively simple type of task in which these laws could be used involves small absolute numerosities where a given number of objects is added to and subtracted from an array of objects (e.g., $2 + 1 - 1$). Inversion tasks of this type were included in the experiment by Starkey and Gelman that was described earlier. As can be seen in Table 8.2, even the 3-year-olds were capable of solving some of the simpler inversion problems. In the majority of inversion problems, overt counting was not used, thus leaving open the possibilities that correct solutions were the result of either covert counting algorithms, the use of an explicitly known inversion property, or memorized number facts. These findings are neutral with respect to the underlying process by which the inversion problems were solved; however, the findings do indicate that even 3-year-olds are capable of providing themselves with data that may be useful in coming to know explicitly the property of inversion.

A somewhat more complicated type of task in which the laws of inversion and compensation could be used involves the use of unknown numerosities. In this type of task, two arrays of objects are usually presented, and the child is given only information about the relative numerosity of the arrays (i.e., information as to whether one array contains more, fewer, or the same number of objects as the other array). Information about relative numerosity can be provided in at least two ways, without providing information about absolute numerosity. The arrays can be screened from view so that the experimenter can simply assert that the relative numerosity of the two arrays is of a particular type (e.g., equal). Alterna-

TABLE 8.2
Mean Proportions of Correct Answers on Inversion Problems

Problems	*3-Year-Olds*	*4-Year-Olds*	*5-Year-Olds*
1 + 1 − 1 and 1 − 1 + 1	.81	.97	1.00
2 + 1 − 1 and 2 − 1 + 1	.62	.88	.91
3 + 1 − 1 and 3 − 1 + 1	.50	.75	.97
4 + 1 − 1 and 4 − 1 + 1	.31	.88	.94
1 + 2 − 2	.44	.88	.94
3 + 2 − 2 and 3 − 2 + 2	.34	.57	.75
[a]2 + 2 − 1 and 2 − 1 + 2	.22	.79	.82

[a]"Incomplete inversion" problems.

tively, arrays can be placed in one-to-one correspondence in a situation in which the child is instructed not to count. Thus, it is possible to present tasks in which the child identifies relative numerosity but not specific numerosities. Inversion tasks of this type have been presented by Smedslund (1966a), Brush (1978), and Cooper et al. (1978), and compensation tasks have been presented by Smedslund and by Cooper et al.

Smedslund's (1966a) study is part of a larger set of studies (Smedslund, 1962, 1966b, 1966c) intended in part to determine the source of difficulty with problems involving combinations of addition and subtraction transformations. Smedslund's task consisted of a series of transformations performed on either or both of two screened, homogeneous arrays that initially contained 16 objects each. Smedslund found that the majority of 5- and 6-year-olds can correctly solve simple inversion and compensation problems where the relative numerosity but not the specific numerosity of the arrays is known. Similar results were obtained from 4- and 5-year-olds on inversion problems by Brush (1978) and from 4- and 5-year-olds but not 3-year-olds on inversion and compensation problems by Cooper et al. (1978).

In the studies by Brush and by Cooper et al., the initial arrays were construced by an "iterative, temporal" one-to-one correspondence procedure. One object was placed in one array at the same time that one object was placed in another array, a procedure that continued until the initial arrays were fully constructed. Brush and Cooper et al. avoided the problem of children responding to misleading length cues in the posttransformation arrays by presenting the arrays in containers. Cooper et al. also presented an inversion problem in which spatial rather than iterative, temporal one-to-one correspondence cues were provided with the initial arrays. The inversion transformation (subtracting and then adding one object) was performed such that misleading spatial cues were present in the posttransformation arrays. It was found that a significant number of children who passed the corresponding inversion problem in which no misleading spatial cues were present in the posttransformation arrays failed the inversion problem in which misleading spatial cues were present in the posttransformation array. These tasks were failed because many children ignored the sequence of transformations and used the misleading length cues to make number judgments as is done by young children in standard number conservation tasks.

In addition to the simple inversion and compensation problems, both Smedslund and Cooper et al. presented incomplete inversion and incomplete compensation problems, incomplete because two objects were initially added (or subtracted) but only one object was subsequently subtracted (or added). These problems are of interest because they bear on the explanation for the findings obtained on simple inversion and compensation tasks.

In the Cooper et al. study, the majority of 5-year-olds gave correct answers throughout the simple and incomplete inversion and compensation problems. Cooper et al. labeled their solutions as "quantitative" solutions, a label reflecting the adequacy and appropriateness of the solutions employed. The 3-year-olds gave incorrect answers on both types of inversion and compensation problems. Their errors indicated that they ignored all previous transformations of the arrays except the last transformation when asked to make number judgments. If array *a* underwent an addition transformation, it was judged most numerous; if array *b* subsequently underwent an addition transformation, it was judged most numerous; or if an array had most recently undergone a subtraction transformation, it was judged least numerous. Cooper et al. refer to this type of solution as "primitive," as it apparently is the earliest type of solution procedure employed on this type of problem.

The majority of the 4-year-olds gave correct answers on the simple inversion and compensation problems but incorrect answers on the incomplete inversion and compensation problems. A significant majority of their incorrect answers were judgments that a complete inversion or complete compensation had occurred when it had not. These "qualitative" solutions were based only on the type of the transformations that were performed and ignored the number of objects used in the transformations. A reanalysis of Smedslund's (1966a) data reveals that the 5- and 6-year-old children in his study made the same type of error as the 4-year-olds in the Cooper et al. study.

Gelman (1972a, 1977) has found that, in tasks somewhat different from those described, 3- and 4-year-olds make the same type of "qualitative" error. In one study (Gelman, 1972a), the children were presented a "magic" task in which an object was surreptitiously subtracted from a three-object array. The children correctly inferred that a subtraction transformation had occurred and were able to cancel the effects of the transformation by adding an object. However, in a second study (Gelman, 1977), two objects were surreptitiously subtracted from a five-object array. The children inferred that a subtraction transformation had occurred, but, though they added in order to cancel the effect of the subtraction transformation, they usually failed initially to add exactly two objects. This suggests insensitivity to the exact number of objects participating in a transformation.

Children giving "qualitative" solutions look as if they are using the following implicit set of rules:

1. If $a = b$ and $c = d$, then $a + c = b + d$.
2. If $a = b$ and $c < d$, then $a + c < b + d$.
3. If $a = b$ and $c > d$, then $a + c > b + d$.
4. If $a < b$ and $c < d$, then $a + c < b + d$.
5. If $a < b$ and $c > d$, then $a + c = b + d$.

Rule 5 describes errors based on "qualitative" solutions. If something akin to this set of rules is used by young children, inadequately detailed information is being used in solving these problems. In some sense, this set of rules contains inversion and compensation laws that fail to specify that the same number of objects must be added or subtracted. Interestingly, the acquisition of these overly general rules and their subsequent restriction appears to occur in tandem for inversion and compensation. Evidence for this was obtained by Cooper et al. (1978). A comparison of the relative difficulty of the simple inversion problems as opposed to the simple compensation problems revealed a significant biconditional relationship between pass/fail performance on these two types of "reversibility." Children either passed both types of problem or failed both types of problem. (Smedslund's data were not reported in a manner permitting this type of comparison.) This close relationship suggests that some common process is involved with inversion and compensation.

The results on one task administered by Starkey (1978) suggest that the development of an understanding of inversion and compensation as applied to concrete arrays of objects occurs over a rather protracted span of years. The absolute numerosity of both the initial arrays and the addend or subtrahend arrays was unknown to the child, but information was given to the child concerning the relative numerosity of the initial arrays and of the addend or subtrahend arrays. In one set of tasks, relative numerosity information was provided by spatial one-to-one correspondence cues (as is done in number conservation tasks), and addition or subtraction transformations were performed such that misleading spatial cues were present in the posttransformation arrays. This task was very difficult for all age groups tested (4- to 6-year-olds); even the majority of the older age group, who could conserve number, used misleading spatial cues in the compensation problems.

Taken together these results demonstrate that children develop inversion and compensation principles to solve inversion and compensation problems and do not rely on counting strategies or perceptual cues for their solution. Further evidence that children develop general inversion and compensation principles is provided by Goth (1980), who found that young children also give "primitive," "qualitative," and "quantitative" solutions to discrete and continuous quantity tasks (length and amount of modeling clay) involving addition and subtraction. The "primitive–qualitative–quantitative" developmental sequence was obtained for both discrete and continuous quantity tasks; however, discrete quantity solutions were more advanced than continuous quantity solutions. It seems unlikely that children would use counting algorithms to judge which of two quantities of clay contains the most.

To summarize, young children have nonperceptual, noncounting procedures for solving certain types of addition and subtraction problems. An abstract set of rules describes the "qualitative" procedures used by young children, and this set of rules contains overly general laws of inversion and compensation that appear to develop in tandem. Further work is required to give a more detailed description of the nonperceptual, noncounting procedures available to young children confronted with certain types of addition and subtraction problems.

PIAGETIAN THEORY

Piaget's work on addition and subtraction consists of three experiments (Piaget, 1941/1952). The aim of the first was to discover whether the child knew that a whole remained constant irrespective of the various compositions of its parts (e.g., $4 + 4 = 1 + 7$). As in number conservation tasks, the arrays in this experiment contained misleading spatial cues. The child's task was to judge whether eight objects divided into two groups of four was numerically equal to eight objects divided into a group of seven and a group of one object. Piaget found that Stage I children (5- and 6-year-olds) erroneously used misleading spatial cues, that Stage II children correctly solved the task only after empirical verification by counting or spatial one-to-one correspondence, and that Stage III children (7-year-olds) correctly solved the task without the use of (overt) empirical verification. Piaget concluded that only stage III children could be granted true addition understanding, because only these children understood that $(4 + 3) + (4 - 3) = 8$.

The second experiment was concerned with the child's ability to construct two equally numerous arrays from two arrays containing unequal numbers of objects (8 versus 14). Piaget found that Stage I children empirically transferred some objects from the more numerous array, but the children only globally compared the newly constructed arrays; they used misleading spatial cues. Stage II children constructed configurations to compare and equate the two arrays, and Stage III children constructed equivalent arrays by means of an a priori decomposition of the arrays (i.e., by means of a deductive solution rather than an empirical trial-and-error solution). Piaget concluded that numerical addition and subtraction are operations only in Stage III children, as only these children are capable of using reversibility by inversion.

The third experiment was concerned with the child's ability to construct two equally numerous arrays from a single array. The response patterns of the Stage I and Stage II children were similar to those for the second experiment. The Stage III children solved the task by use of spatial one-to-one correspondence in constructing the two arrays.

Some of Piaget's general conclusions are that an operational understanding of addition–subtraction is not present unless the child can perform enumeration and colligation, one as a function of the other. The argument is that the union of these two processes into a single whole is necessary for understanding one-to-one correspondence, and one-to-one correspondence is the link between enumeration and colligation. In order for addition–subtraction to be operational, the child must recognize that colligation and enumeration are necessarily the inverse of one another. (By counting the elements in one array, one infers the numerosity of another array without counting if the arrays are in one-to-one correspondence.)

Relation of Inversion and Compensation to Number Conservation

Inversion and compensation are types of reversibility Piaget specifically mentions as being concrete operational competencies. Inasmuch as, according to Piaget, number conservation requires reversibility, it seems reasonable to hypothesize an empirical relationship between the ability to conserve number and the ability to solve some types of inversion and compensation problems. Smedslund (1966a) found that his number conservation problem was more difficult to solve than the inversion and compensation problems we described earlier; however, Smedslund reported a problem with his conservation procedure that apparently resulted from the absence of salient one-to-one correspondence cues in the initial arrays. The specific problem was that the children correctly choose the array that was numerically equal to the initial array; however, when questioned about the relative numerosity of the chosen and standard arrays, several children changed their minds and decided that the arrays were unequal. We agree with Smedslund that this problem is sufficiently serious to warrant caution in interpreting his conservation data.

Cooper et al. (1978) also included number conservation tasks in their study of inversion and compensation. It was found that inversion and compensation tasks containing no misleading length cues were significantly easier than standard number conservation tasks. Interestingly, though, a significant biconditional relationship was not found to exist between performance on the inversion and conservation tasks; that is, it was not the case that a significant proportion of children who solved the inversion problem also solved the conservation problem, and vice versa. Several children passed inversion but not conservation, and several children displayed the converse performance pattern. Thus, a close empirical relationship of the type one might expect from Piagetian theory was not found for conservation and inversion–compensation.

Starkey (1978) included number conservation tasks in the previously described study of a rather difficult type of compensation. It was found that number conservation was significantly easier than the compensation problems. Thus, the consistent finding across the studies of Smedslund, Cooper et al., and Starkey is that a biconditional relationship has not been found between number conservation and inversion–compensation. The relative difficulty of number conservation and inversion–compensation depends on the specific type of conservation, inversion, and compensation tasks presented.

A Critique of Piaget's Theory of Number

We now examine the adequacy of Piaget's (1941/1952) theory of the child's conception of number to account for the findings reviewed up to this point. Piaget recognized that young children can give correct answers to number-fact problems as well as count, but, as indicated by the following quotations, he discounts the importance of number fact and counting data obtained from "pre-operational" children:

> It is true that even children who are still at the earlier stages can be taught to repeat formulae such as $2 + 2 = 4$; $2 + 3 = 5$; $2 + 4 = 6$, etc., but there is no true assimilation until the child is capable of seeing that six is a totality, containing two and four as parts, and of grouping the various possible combinations in additive compositions. When these conditions do not obtain, addition as an operation is not understood. The child at the first stage perceives that when $4 + 4$ becomes $7 + 1$, one of the subsets increases, but this intuition becomes addition only when this increase is compensated by a decrease $(4 + 3) + (4 - 3) = 8$. It is this interdependence of direct and inverse operations that we shall now examine in relation to a single, typical example [p. 190].

> counting aloud appears to have little influence on the belief in the equivalence of two sets as a result of one-to-one correspondence. We have already frequently had occasion, in earlier sections, to point out that there is no connection between the acquired ability to count and the actual operations of which the child is capable [p. 61].

> If the child has not yet reached a certain level of understanding which characterizes the beginning of the third stage, counting aloud has no effect on the mechanism of numerical thought [p. 63].

It seems that Piagetian theory must account for number development in children up to approximately 7 years of age without allowing the possibility that counting, counting algorithms (by implication), and number-fact information contribute to early development. Furthermore, these disallowed abilities and knowledge do not even have the status of being precursors of later numerical abilities or knowledge.

One criticism of the Piagetian view is the arbitrary manner in which data on the ability of young children to solve number-fact problems, to count, and to use counting algorithms are dismissed as rote processes, the products of which have no numerical meaning or utility to the child. Piaget asserted as much, but never empirically demonstrated or proved the assertion. A related problem has to do with the data described earlier on self-generated counting algorithms. If counting is a rote process, how could adequate (i.e., accurate and valid) counting algorithms be invented by the child, how could young children use information obtained by counting to make more/less judgments and to counter the effect of a transformation, and how could young children solve number fact problems such as $14 + 1 =$ ______ or $15 - 1 =$ ______ that they have never memorized? These findings indicate that Piaget's assertions are incorrect.

A second type of criticism of Piagetian theory has to do with the general stage transitions and states proposed in the theory. The existence of horizontal decalage across content areas (e.g., discontinuous versus continuous quantity) is now recognized in Piagetian theory and explained in terms of the relative resistances of particular content areas to cognitive operations (e.g., see Piaget's introduction to Laurendeau & Pinard, 1970); thus findings such as the horizontal decalage observed by Goth (1980) would not be a problem for the theory. Other findings such as inversion in "preoperational" children, are more of a problem, as the theory does maintain that within a particular content area or cognitive domain children either do or do not possess the structures characterizing a given stage of development (e.g., the set of concrete operations). Some of the findings discussed earlier do not support this aspect of Piaget's theory, because they indicate that, within the number domain, the development of "concrete operational" processes occurs in a piecemeal manner rather than in a holistic, all-or-none manner. The data on inversion and compensation in the absence of misleading spatial cues (Cooper et al., 1978; Smedslund, 1966a) are of this type.

Add to these results the fact that it is easy to change preschoolers' nonconservation performance to conservation (Markman, 1979). Clearly there are problems with the view of abrupt stage changes. Thus, at least two criticisms apply to the Piagetian theory of number development. Piaget errs when he dismisses, as irrelevant, data on early counting, number-fact solutions, and counting algorithms. Second, the view that rather clearly defined stages and sharp stage transitions exist within particular cognitive domains needs modification.

AN ALTERNATIVE VIEW

Data from a variety of sources are converging on the view that number forms a natural cognitive domain. Children naturally are able to separate number from other basic categories, and cultural transmission is not necessary for separating number from domains such as space, time, or causality. It appears that coming to know about number is much like coming to know about language (Gelman, 1979). The ability to learn language is rule governed; the ability to learn to count verbally is rule governed (Gelman & Gallistel, 1978). Language learning depends on a supporting environment, but it does not depend on schooling; counting also appears prior to schooling. Language is a cultural universal, as are some number abilities; an especially clear case can be made for counting (Ginsburg, 1978, and in this volume; Menninger, 1969; Saxe, 1979; Zaslavsky, 1973). During the course of language acquisition, children spontaneously rehearse, generate rules, and self-correct; during the course of early number development, children spontaneously generate counting algorithms and self-correct some counting errors (See Fuson in this volume; Gelman & Gallistel, 1978; Groen & Resnick, 1977). Infants attend to, discriminate, and represent speech sounds; infants also attend to, discriminate, and represent particular small numerosities such as "twoness" (Cooper, Starkey, & Dannemiller, in preparation; Starkey & Cooper, 1980; Starkey, Spelke, & Gelman, 1980). Retardates have some language ability and some number ability. And, skill at counting is diagnostic regarding the retardate's ability to solve simple addition and subtraction tasks (Starkey & Gelman, in preparation).

These findings, taken as a whole, support the view that some number abilities are natural human abilities in the same sense that some language abilities are natural human abilities. As such, this view (in contrast with the Piagetian view) emphasizes the importance of counting, counting algorithms, and solutions to basic arithmetic problems as relevant and important aspects of the young child's understanding of number. Counting, or, more generally, the ability to abstract precise numerosities across discrete objects, appears to be a central and basic ability in the number domain. Numerical abstraction is used not only to establish the numerosity of static (i.e., untransformed) arrays of objects, but, perhaps more importantly for development, to establish the effects of various types of transformations on numerosity (Klahr & Wallace, 1976). Thus, numerical abstraction may contribute to the development of an understanding of conservation and of the basic laws of arithmetic.

Our view not only differs from the Piagetian view concerning the relevance of data on early number abilities such as counting, but also differs as to whether clearly defined stages and sharp stage transitions exist within the number domain. We suggest that young children possess localized rather than domain-wide competencies and knowledge. For example, at some point in development the inversion law is recognized to apply in tasks containing no misleading spatial cues; at a later point it is recognized to apply also to a wider range of tasks such as tasks containing misleading spatial cues or tasks involving fractions. At one point children conserve number when collection terms are used; at a later point they also conserve when class terms are used. The range of tasks to which a particular arithmetic law (e.g., compensation) is recognized as applicable is not necessarily identical to the range for some other particular arithmetic law (e.g., commutativity). Thus it is proposed that an early understanding of arithmetic is local to particular principles, and further, the understanding of these principles proceeds through more and more complex levels. This contrasts with the Piagetian view of a stage of concrete operations where a diverse set of quantitative principles come to be understood simultaneously.

REFERENCES

Beilin, H. Cognitive capacities of young children: A replication. *Science,* 1968, *162,* 920-921.

Bjonerud, C. D. Arithmetic concepts possessed by the preschool child. *Arithmetic Teacher,* 1960, *7,* 347-350.

Brush, L. R. Preschool children's knowledge of addition and subtraction. *Journal for Research in Mathematics Education,* 1978, *9,* 44-54.

Buckingham, B. R. Teaching addition and subtraction facts together or separately. *Educational Research Bulletin,* 1927, *6,* 228-229, 240-242.

Clapp, F. L. The number combinations: Their relative difficulty. *Bureau of Educational Research Bulletin,* No. 2, 1924.

Cooper, R. G., Starkey, P., Blevins, B., Goth, P., & Leitner, E. *Number development: Addition and subtraction.* Paper presented at the meeting of the Jean Piaget Society, Philadelphia, May 1978.

Cooper, R. G., Starkey, P., & Dannemiller, J. *Numerical representation in human infants.* In preparation.

Dutton, W. H. Growth in number readiness in kindergarten children. *Arithmetic Teacher,* 1963, *10,* 251-255.

Gelman, R. Logical capacity of very young children: Number invariance rules. *Child Development,* 1972, *43,* 75-90. (a)

Gelman, R. The nature and development of early number concepts. In H. W. Reese (Ed.), *Advances in child development and behavior* (Vol. 7). New York: Academic Press, 1972. (b)

Gelman, R. How young children reason about small numbers. In N. J. Castellan, D. B. Pisoni, & G. R. Potts (Eds.), *Cognitive theory* (Vol. 2). Hillsdale, N.J.: Lawrence Erlbaum Associates, 1977.

Gelman, R. *The nature of numerical abilities in preschoolers.* Invited address given at the annual meeting of the American Educational Research Association, San Francisco, April 1979.

Gelman, R., & Gallistel, C. R. *The child's understanding of number.* Cambridge, Mass.: Harvard University Press, 1978.

Ginsburg, H. Poor children, African mathematics, and the problem of schooling. *Educational Research Quarterly,* 1978, *2*(4), 26-44.

Goth, P. *The development of addition-subtraction knowledge and its relation to conservation in young elementary school children.* Unpublished doctoral dissertation, University of Texas at Austin, 1980.

Groen, G. J., & Parkman, J. M. A chronometric analysis of simple addition. *Psychological Review,* 1972, *79,* 329-343.

Groen, G. J., & Poll, M. Subtraction and the solution of open sentence problems. *Journal of Experimental Child Psychology,* 1973, *16,* 292-302.

Groen, G. J., & Resnick, L. B. Can preschool children invent addition algorithms? *Journal of Educational Psychology,* 1977, *69,* 645-652.

Ilg, F., & Ames, L. B. Developmental trends in arithmetic. *Journal of Genetic Psychology,* 1951, *79,* 3-28.

Klahr, D., & Wallace, J. G. *Cognitive development, an information-processing view*. Hillsdale, N.J.: Lawrence Erlbaum Associates, 1976.

Knight, F. B., & Behrens, M. S. *The learning of the 100 addition combinations and the 100 subtraction combinations*. New York: Longmans, Green, 1928.

Laurendeau, M., & Pinard, A. *The development of the concept of space in the child*. New York: International Universities Press, 1970.

Markman, E. Classes and collections: Conceptual organization and numerical abilities. *Cognitive Psychology*, 1979, *11*, 395–411.

McLaughlin, K. L. *A study of number ability in children of ages three to six*. Doctoral dissertation, University of Chicago: University of Chicago Press, 1935.

Mehler, J., & Bever, T. G. Cognitive capacity of very young children. *Science*, 1967, *158*, 141–142.

Menninger, K. *Number words and number symbols, a cultural history of numbers*. Cambridge, Mass.: MIT Press, 1969.

Mott, S. M. Number concepts of small children. *Mathematics Teacher*, 1945, *38*, 291–301.

Piaget, J. [*The child's conception of number*] (C. Gattegno & F. M. Hodgson, trans.). London: Routledge & Kegan Paul, 1952 (Originally published, 1941).

Richard, E. E. S. *An inventory of the number knowledge of beginning first grade school children, based on the performance of selected number tasks*. Unpublished doctoral dissertation, Indiana University and Indiana State University, 1964.

Saxe, G. B. Children's counting: The early formation of numerical symbols. *New Directions for Child Development: Early Symbolization*, 1979, *3*, 73–84.

Smedslund, J. The acquisition of conservation of substance and weight in children, VII. Conservation of discontinuous quantity and the operations of adding and taking away. *Scandinavian Journal of Psychology*, 1962, *3*, 69–77.

Smedslund, J. Microanalysis of concrete reasoning, I. The difficulty of some combinations of addition and subtraction of one unit. *Scandinavian Journal of Psychology*, 1966, *7*, 145–156. (a)

Smedslund, J. Microanalysis of concrete reasoning, II. The effect of number of transformations and non-redundant elements and of some variations in procedure. *Scandinavian Journal of Psychology*, 1966, *7*, 157–163. (b)

Smedslund, J. Microanalysis of concrete reasoning, III. Theoretical overview. *Scandinavian Journal of Psychology*, 1966, *7*, 164–167. (c)

Starkey, P. *Number development in young children: Conservation, addition and subtraction*. Unpublished doctoral dissertation, University of Texas at Austin, 1978.

Starkey, P., & Cooper, R. G. Perception of numbers by human infants. *Science*, 1980, *210*, 1033–1035.

Starkey, P., & Gelman, R. *Addition and subtraction abilities of normal and retarded children*. In preparation.

Starkey, P., Spelke, E., & Gelman, R. *Number competence in infants: Sensitivity to numeric invariance and numeric change*. Paper presented at the meeting of the International Conference on Infant Studies, New Haven, March 1980.

Svenson, O. Analysis of time required by children for simple additions. *Acta Psychologica*, 1975, *39*, 289–302.

Svenson, O., & Broquist, S. Strategies for solving simple addition problems, a comparison of normal and subnormal children. *Scandinavian Journal of Psychology*, 1975, *16*, 143–151.

Svenson, O., Hendenborg, M., & Lingman, L. On children's heuristics for solving simple additions. *Scandinavian Journal of Educational Research*, 1976, *20*, 161–173.

Thyne, J. M. *Patterns of error in the addition number facts*. London: University of London Press, 1954.

Wheeler, L. R. A comparative study of the difficulty of the 100 addition combinations. *Pedagogical Seminary and Journal of Genetic Psychology*, 1939, *54*, 295–312.

Williams, A. H. Mathematical concepts, skills and abilities of kindergarten entrants. *Arithmetic Teacher,* 1965, *12,* 261-268.

Woods, S. S., Resnick, L. B., & Groen, G. J. An experimental test of five process models for subtraction. *Journal of Educatonal Psychology,* 1975, *67,* 17-21.

Woody, C. The arithmetical backgrounds of young children. *Journal of Educational Research,* 1931, *24,* 188-201.

Zaslavsky, C. *Africa counts, number and pattern in African culture.* Boston, Mass.: Prindle, Weber, & Schmidt, 1973.

9 Towards a Generative Theory of "Bugs"

John Seely Brown
Kurt VanLehn
Cognitive and Instructional Sciences
Xerox Palo Alto Research Center

During the past few years our group has been working on computer systems for diagnosing systematic student errors. These diagnostic systems, BUGGY and more recently DEBUGGY, have been used to analyze thousands of students (Brown & Burton, 1978; Burton, 1981) and have enabled us to construct an extensive catalogue of precisely defined bugs for place-value subtraction. These procedural definitions constitute an extensional, ad hoc theory of bugs. The theory is extensional in that it explains a student's errors and misconceptions as a combination of some set of prespecified, primitive bugs. It is ad hoc (by design!) in that the language we used to describe the bugs was as open-ended and as powerful as we could find. Although we could have started out using a language restricted by psychological principles, we chose to use a powerful descriptive technique that facilitated our ability to express and experiment with any new bug that was conjectured to exist in the data.

In this chapter we describe our current efforts to move from our ad hoc extensional theory to a generative theory, one that is capable of explaining why we found the bugs that we did and not other ones, one that is capable of explaining how bugs are caused, and most importantly, one that is capable of predicting what bugs will exist for procedural skills we have not yet analyzed. There are several benefits of such a theory. We could automatically generate a list of bugs for a new skill and add these new bugs to our current diagnostic system, thereby creating a diagnostic system tailored to the new skill. We could attack the issue of the remediation of bugs with more than just a knowledge of what bugs a student has, because such a theory would provide a plausible basis for understanding why the student had those bugs. Such an understanding could also help us design learning environments that might inhibit those bugs in the first place. Finally, in terms of cognitive research, such a theory would provide insights into knowledge representations and cognitive mechanisms that defy direct observation.

The Concept of a Bug in Arithmetic

An arithmetic procedure is not a single, simple procedure. It has internal structure. Multidigit subtraction, for example, is composed of the subprocedures of single-digit subtraction, borrowing, and borrowing across zero. But even this structure is too coarse a description of subtraction for our purposes. There is a finer structuring of arithmetic procedures that appears to play an important role in students' arithmetic behavior, their erroneous behavior in particular.

This fine structure can be seen when a student's errors are carefully examined. Once we look beyond what kinds of exercises the student misses (which usually provides the data for a skill analysis) and look at the actual answers given, we find in many cases that these answers can be precisely predicted by hypothetically computing the answers to the given problems using a procedure that is a small perturbation in the fine structure of the correct procedure. Such perturbations serve as a precise structural description of the errors. From this perspective, errors are the symptoms manifested by a student following a variant of the correct procedure. Thus they are often seen to be extraordinarily systematic, once the right variant to the correct procedure has been diagnosed.

To form such predictions, the perturbation itself must be described exactly. Precisely defined erroneous variations of a procedure are known as "bugs" in the computer world and we have adopted that term here. That is, we are not only interested in what subskills the student has not yet mastered, but we are also interested in what, if anything, stands in the place of those unmastered subskills, that is, what bugs the student has. For example, consider the seven problems below:

206	80	183	702	513	800	3005
−38	−4	−95	−11	−268	−168	−28
138	56	88	691	245	662	2027

One could vaguely describe these problems as coming from a student having trouble with borrowing, especially in the presence of zeros (i.e., the skills of borrowing or decrementing in the context of zeros have not been mastered). But such a description of the student's trouble is hardly complete. It allows one to make a good guess about what new problems the student is likely to get wrong, but it does not allow prediction of the actual answers. The description is just not precise or complete enough to simulate the student's actual behavior on a new problem.

Examples of Some Subtraction Bugs

Most of the common bugs are well known to educators. Here are three examples:

1. Smaller-From-Larger. Instead of borrowing, the student takes the top digit, which is smaller, from the bottom one, which is larger.
2. Borrow-From-Zero. When borrowing from a zero, the student changes the zero to a nine, but doesn't go on to borrow from the next digit to the left.
3. Diff-0 $- N = N$. When the top digit is zero, the student writes down the bottom digit as the column answer.

Borrow-From-Zero illustrates that not all bugs introduce new subprocedures into the procedure. In this bug, half the borrowing across zero subprocedure is missing, but no additional subprocedures have been added.

Following are some bugs that occur much less frequently. Included in their description are a few "bug stories," which are conjectures about what might be going through a student's head as he or she invents the bug. Although we have no basis for these stories, they help focus the search for principled explanations of the bugs.

1. Add-Decrement-Instead-Of-Borrow-From-Zero. Instead of borrowing across zero, the student changes the zero to a one. A story for this bug is that during borrowing the student is supposed to subtract one from a digit, but can't because the digit is a zero. The next best thing is to add one to it because that's what is done in a similar situation, namely carrying in addition.
2. Borrow-Across-Zero. When borrowing across one (or more) zeros, the student does not change the zero(s) to nine(s), although the first nonzero digit is decremented correctly. The student knows to decrement a number and because he can't decrement zero he keeps moving left trying to find a number that can be decremented. The first such number found is decremented.
3. Forget-Borrow-Over-Blanks. During borrowing, if the digit to be decremented is over a blank, the student doesn't decrement. One story for this bug is that the student is using the Equal Additions method of subtraction, which involves incrementing the bottom digit during a borrow instead of decrementing the top digit. A blank can't be incremented because there is no number.

Other bugs are discussed during the course of the chapter.

The Data Base of Observed Bugs

The development of a generative theory of bugs would be impossible without an accurate and nearly complete description of what bugs exist, especially considering that part of the task of a generative theory is to explain why certain bugs don't exist as well as to explain why other bugs do exist. The design, accumulation, and verification of such a data base of bugs presents many difficult research problems. Some of these have been solved (Brown & Burton, 1978; Burton, 1981) and others are still under investigation.

A question that plagues all such empirical investigations concerns the criteria for claiming that a student has a particular bug. There are several options. One might say that a student has a bug only if almost every answer on the test is predicted by the bug. Another possibility is that some answers on the test can be predicted by the bug. Or weaker still, one might require that some proportion of the columns of an answer on the test be generated by the bug. Yet still weaker is the standard skill notion that a "bug" need only predict which problems the student would get wrong but not necessarily what his answers would actually be. We chose the first, most stringent criterion and according to the initial results of ongoing study by our group, over one-third of the student sample has bugs that meet this criterion!

However, the percentage of students having bugs according to this strict definition is irrelevant to the kind of generative theory we have in mind. Rather, what is important is whether a bug exists in the population at large. Indeed, when we have been able to obtain the students' answer sheets, it is sometimes possible to see in the scratch marks a bug that has been obscured by "noise" such as misrecollection of basic number facts. Such bugs are said to exist even though it is not strictly true that almost all the student's answers are predicted by the bug.

It is very frequently the case that a student appears to have more than one bug at the same time. Thus, for example, the student might have both Borrow-From-Zero and Diff-$N - 0 = 0$. Thus, the data base is actually structured as a certain number of primitive bugs (currently 88), and information about how to combine them into compound bugs. The combination of primitive bugs into compound bugs is not usually as straightforward as the foregoing example would lead one to believe. Consequently, instead of the 89^4 compound bugs that are logically possible (we only consider compounds of four or fewer primitive bugs), something like 3000 bugs are in fact possible (see Burton 1981 for details). Of these 3000 bugs, roughly 300 of them have been observed!

In Table 9.1 we provide a description and examples of other primitive bugs that are mentioned in later sections.

There are several different subtraction procedures taught in different parts of the world. Moreover, even in the United States, several variations of the "standard" procedure are taught, differing particularly with regard to the use of scratch marks. In a truly global study, these alternative versions of the correct underlying procedure would be equally important, and so the data base allows various alternative algorithms. However, because our two largest sample populations (1500 and 1000) were drawn from school systems that taught similar subtraction procedures, most of the following theory is couched as if there were only one correct method of subtraction.

TABLE 9.1
Characterization of Sample Subtraction Bugs

Always-Borrow-Left

The student borrows from the leftmost digit instead of borrowing from the digit immediately to the left (733 − 216 = 427).

Blank-Instead-Of-Borrow

When a borrow is needed the student simply skips the column and goes on to the next (425 − 283 = 22).

Blank-Instead-Of-Borrow-From-Zero

When a borrow from zero is needed the student skips the column and goes on to the next (507 − 241 = 36).

Borrow-No-Decrement-Except-Last

Decrements only in last column of the problem (6262 − 4444 = 1828).

Borrow-Won't-Recurse

Instead of borrowing across a 0, the student stops doing the exercise (8035 − 2662 = 3).

Doesn't-Borrow

The student stops doing the exercise when a borrow is required (833 − 262 = 1).

Don't-Decrement-Zero

When borrowing across a 0, the student changes the 0 to 10 instead of 9 (506 − 318 = 198).

Smaller-From-Larger-Instead-Of-Borrow-From-Zero

The student does not borrow across 0. Instead he or she will subtract the smaller from the larger digit.

$$\begin{array}{r} 306 \\ -\ \ 8 \\ \hline 302 \end{array} \qquad \begin{array}{r} 306 \\ -148 \\ \hline 162 \end{array}$$

Stops-Borrow-At-Zero

Instead of borrowing across a 0, the student adds 10 to the column he or she is doing but doesn't decrement from a column to the left (404 − 22 = 227).

Stutter-Subtract

When there are blanks in the bottom number, the student subtracts the leftmost digit of the bottom number in every column that has a blank (4369 − 22 = 2147).

Zero-After-Borrow

When a column requires a borrow, the student decrements correctly but writes 0 as the answer (65 − 48 = 10).

Zero-Instead-Of-Borrow

The student doesn't borrow; he or she writes 0 as the answer instead (42 − 16 = 30).

Zero-Instead-Of-Borrow-From-Zero

The student won't borrow if he or she has to borrow across 0. Instead he or she will write 0 as the answer to the column requiring the borrow.

$$\begin{array}{r} 702 \\ -\ \ 8 \\ \hline 700 \end{array} \qquad \begin{array}{r} 702 \\ -348 \\ \hline 360 \end{array}$$

Bug Stability

Given the fine-grain analysis provided by the extensional theory a phenomenon has emerged that has just begun to get some empirical support. Using some of the BUGGY ideas, Hetzel (1979) detected that some students diagnosed as having a particular bug one day would appear to have an equally systematic but different bug a few days later (this often happened over a weekend where there was no intervening instruction). Since then, we have discovered that many educators believe that the systematic errors they observe in a given student often spontaneously disappear and then reappear at later times. Identifying those bugs that can be self-correcting or that inherently lead to slightly different bug formations is a difficult empirical task that is currently underway. Although strong empirical support for even the hypothesis that some bugs are inherently unstable is yet to be obtained, we would hope that our generative theory of bugs might be able to make predictions on what bugs are the most likely ones to be stable or unstable. At the end of this chapter we show how such predictions are forthcoming.

THE FORM OF THE GENERATIVE THEORY

Our previous work provided both a framework for precisely describing variants (or bugs) of a correct algorithm and a powerful diagnostic tool that we used to sieve large amounts of student data in search of still unaccounted-for errors, which could then be analyzed by hand and if needed, added to our data base of bugs. Now that several thousand student tests have been analyzed we have reached a stage where our data base is converging. We are able to account for a substantial number of student errors and only a small number of new bugs are being discovered. This rather extensive data base of bugs now enables a much deeper question to be investigated—namely, what is the cause of these bugs and why do just they occur and not others?—whereas our earlier effort explained a student's errors as symptoms manifested by bugs in terms of a set of principles that perturb or transform a procedural skill into all its possible buggy variants. We shall call the set of principles and the process that interprets them a generative theory of bugs. Its challenge is twofold. It must generate all the known or expected bugs for a particular skill and it must generate no others.

The theory is motivated by the belief that when a student has unsuccessfully applied a procedure to a given problem, he or she will attempt a repair. Suppose he or she is missing a fragment (subprocedure) of some correct procedural skill, either because he or she never learned the subprocedure or maybe forgot it. Because the missing fragment must have had a purpose, attempting to follow the impoverished procedure rigorously will often lead to an impasse. That is a situation in which some current step of the procedure dictates a primitive action that cannot be carried out, usually because one of its preconditions or input/

output constraints has been violated. For example, an attempt to decrement a zero will lead to an impasse. When a constraint gets violated the student, unlike a typical computer program, is not apt to just quit. Instead he or she will often be inventive, invoking problem-solving skills in an attempt to repair the impasse and continuing to execute the procedure, albeit in a potentially erroneous way. We believe that many bugs can best be explained as patches derived from repairing a procedure that has encountered an impasse while solving a particular problem.

Our generative theory is based on the aforementioned considerations. It postulates that there exists a set of incomplete versions of a given procedure, which are characterized by applying a set of Deletion Principles (or operators) to its complete form. It then postulates that when an incomplete version encounters an impasse as determined by a set of precondition violations for a primitive action contained in the procedural skill, a set of repairs is performed by a Generate-and-Test problem solver. The set of all possible repairs is characterized by a set of repair rules in conjunction with a Tester that can reject or filter out certain proposed repairs based on a set of Critics.

Figure 9.1 details the major constituents of the process model that orchestrates the foregoing principles and shows how they interact to produce a set of bugs. These constituents are:

1. *A representation of the given procedural skill.* In determining this, several issues must be addressed. One concerns the representation language and its associated interpreter. Another concerns the actual structural decomposition of

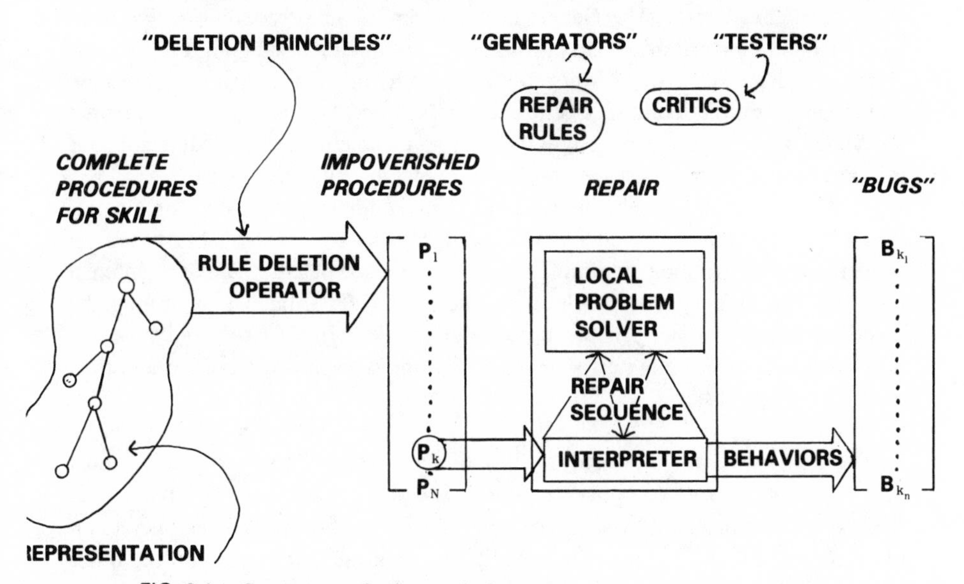

FIG. 9.1. Components of naive repair theory.

the skill that is to be embedded in the chosen representation language. The same procedural skill or method can often be decomposed in more than one way, which, in turn, can have subtle theoretical ramifications.

2. *A set of principles for deleting (or editing) fragments of the correct procedure.* These principles will characterize what parts of the original skill can be deleted, thereby reflecting what parts of the procedure might become inaccessible in the long-term memory or may never have been learned (given the circumstances of our testing, it is often the case that students receive problems requiring subprocedures that they have not yet been taught). For example, the simplest principle might assert that any step of a procedure can be deleted; other principles might restrict the deletions to reflect a possible learning sequence of the procedure.

In understanding our use of deletion operators it is critical to realize that we are not claiming that a student necessarily knew the correct procedure and then forgot part of it, namely the deleted part. We are using the idea of deletion as a means to characterize formally (as opposed to explain) the set of incomplete or otherwise defective procedures that are candidates for use and possible repair. A learning theory is being constructed that can generate that same set of candidates by simulating a student's miscomprehension of some of the examples used in a teaching sequence. In short, we use deletion as a precise way to generate a certain set of procedures, realizing that a deep explanation for this set may be found in theories of forgetting and mislearning. In this chapter we do not give a precise definition of the deletion operators or the representation language that they act on (see VanLehn, 1980).

3. *The Generator.* The Generator can examine the preconditions that have been violated on a primitive and propose explicit repairs based on a set of Repair Rules. Our later discussion of this component circumvents control issues of how one repair rule might be initially chosen over another. Instead, we focus on what the actual repair rules are and claim that any applicable repair rule that is not later rejected by the Tester must generate an acceptable bug.

4. *The Tester.* Closely allied to the Generator is the Tester (or Critic), which filters out those repairs that it considers to be unreasonable based on properties of the solution stemming from the proposed patch. Again, our interest here is on the precise set of filtering conditions or "Critics" constituting the Tester and not so much on the process of invoking the Critics and how the necessary backtracking is performed when an initial repair rule is tried and then rejected by the Tester.

There are several noteworthy points to the form of this theory. The most important concerns its composite nature. We could have tried to account for all the known bugs in a skill by searching for a set of transformations that operate on the skill and directly produce all and only those bugs. Our theory, on the other hand, involves two stages. The first stage edits the skill as characterized by a set of deletion principles. The resulting edited procedures are not intended to be the sought-after bugs (although some of them may be). Instead, each possible edit generates a procedural variant that when followed (or executed) will often lead to an impasse that invokes stage two, the repair process. This second stage uses a set of repairs to fix the existing impasse. It is the set of all possible repairs to all possible interrupts that is meant to characterize the set of all possible bugs.

Methodology

The goal of a generative theory is to generate all the known bugs, predict some new bugs, but avoid predicting old, improbable bugs. Because a primary concern is to provide a principled account as opposed to an ad hoc account of a set of bugs and to use these principles to predict bugs for skills yet to be analyzed, we invoke as little problem-solving machinery as possible to account for the data. We fully recognize that there exist much more powerful problem-solving models that may, in fact, better capture what a student is actually thinking while inventing a patch to an encountered impasse. We also utilize as little of an actual process model as is possible and instead proceed under the assumption that if a rule is applicable it will be used. The trouble with invoking more problem-solving machinery as a process model is that it is hard to get a crisp boundary on what bugs the model will generate because, for example, it is never certain what control strategies a student might be using to select and specialize his or her repair rules. We sidestep such issues and see just how far we can get with repair rules that apply to any impasses encountered while following the procedure. Although the methodology of characterizing a phenomenon by pushing a simple technique to the extreme (even when that means using it in contorted ways) before invoking more complex techniques can sometimes produce psychologically questionable models, we find that it can also produce models that are highly understandable and predictive.

We have also adopted the methodological principle that each piece of information in the representation of a procedure must be usable in the correct solution of at least one relevant problem. This principle rules out the representation of bugs as "dead code," or information that is accessed only in the case of a deletion. With no principles governing the presence of dead code, allowing it would mean that the explanation for a bug would not be completely contained within the theory, a situation we would like to avoid.

REPAIR GENERATION

We believe that a student following a procedure that specifies that a particular primitive is now to be executed but can't be, for whatever reason, is apt to invent some repair to circumvent the dilemma. For example, suppose he or she is trying to perform a column subtract with a larger number from a smaller number and can't because there is no appropriate entry in his or her facts table (or because he or she believes he or she can't). What might he or she do? One obvious repair might be to skip trying to execute that primitive action and move on to the next step of the procedure. Another repair might be to simply quit doing the problem. Yet another repair might be to switch the arguments to his or her facts-table look-up procedure—that is, if it doesn't work with one sequence of arguments, try swapping them around. And a last example might involve resorting to invoking the counting-based subtraction procedure originally used to generate or understand the facts table. Of course, the counting procedure eventually generates its own impasse or precondition violation due to the first input being larger than the second. But then suppose the student recursively tries to repair this impasse with, say, a quit repair. For example, if the "count-down" procedure is used, the wrong counter becomes zero first. Usually, this counter is nonzero when the other counter zeros out. Suppose the response to this precondition violation is simply to quit, returning to the usual counter, which in this case is zero. The overall effect of this repair sequence is to write a zero into the column's answer.

A Detailed Example

We have found that the optimal structure for enabling repairs to correct subtraction contains the following rules (VanLehn, 1980):

1. If the goal is to do a subtraction problem, then Subtract each Column separately in succession from right to left.
2. If the goal is to Subtract a Column, then take the bottom digit from the top digit and write their difference as the answer. (For future reference this operation of taking the bottom digit from the top will be called "Diff.")
3. If the goal is to Subtract a Column, and the bottom of the column is blank, then write the top digit of the column as the answer.
4. If the goal is to Subtract a Column, and the top digit is smaller than the bottom digit, then Borrow.
5. If the goal is to Borrow, then add 10 to the top digit in this column.
6. If the goal is to Borrow, then Try To Decrement the next column to the left.
7. If the goal is to Try To Decrement a column, then subtract one from the top digit and write the result above it.
8. If the goal is to Try To Decrement a column, and the top is a zero, then Borrow Across Zero.
9. If the goal is to Borrow Across Zero, then write a nine in place of the top digit.
10. If the goal is to Borrow Across Zero, then Try to Decrement the next column to the left.

(Some of the detail has been suppressed from these rules for legibility.) If Rule 4 is deleted, then Diff will be executed on every column, including larger from smaller (LFS) columns. LFS columns violate a precondition of Diff, namely that the first input number be larger than the second input number. Four bugs can be generated by repairing this violation in different ways. The most natural repair to primitive action whose precondition has been violated is simply to "skip" it, generating Blank-Instead-Of-Borrow. Because Diff is simply skipped when its precondition is violated, the bug causes a failure to write an answer in an LFS column. If the "quit" repair is used, then the bug is Doesn't-Borrow, because the problem is given up as soon as a LFS column is encountered. The "switch-arguments" repair is to swap inputs when they are of the same data type (as they are in this case). When this repair is used, the well-known Smaller-From-Larger bug results, wherein the student takes the absolute difference of each column's numbers. The Zero-Instead-Of-Borrow bug is generated by forgetting about the facts table and reverting to the counting procedure that underlies it. We will call the repair "dememoize," which comes from the computer programming technique of "memoizing" a function by replacing it with a table that pairs its inputs with the outputs it would generate if it had been run. We can summarize these bugs as:

Repairs	*Bugs*
Skip	Blank-Instead-Of-Borrow
Quit	Doesn't-Borrow
Switch Arguments	Smaller-From-Larger
Dememoize	Zero-Instead-Of-Borrow

These four repairs can be used in conjunction with the deletion of Rule 5 to generate four more bugs. When Rule 5 is deleted, the procedure does the decrement part of borrow correctly but fails to add 10 to the top digit of the column borrowed into. Hence, when control returns from Rule 4, and Rule 2 is run, Diff is entered with the column in its original LFS state. Hence, the familiar precondition is violated, and the same four repairs generate four new bugs (one of which has not been observed, and is a prediction made by the theory):

Repairs	*Bugs*
Skip	Blank-With-Borrow (this bug has not been observed, yet)
Quit	Borrow-Won't-Recurse
Switch Arguments	Smaller-From-Larger-With-Borrow
Dememoize	Zero-After-Borrow

Another group of four bugs can be generated by deleting Rule 8. This means that when the procedure is given a problem that requires borrowing across zero,

Rule 7 will run instead of the missing Rule 8, and Decrement will be called with zero as its input. This violates its precondition. The four bugs we're interested in come from supposing that this violation is repaired by the repair "Backup." Backup is a well known strategy in problem solving: One "backs up" control to the last point where a choice was made. In this context, that means moving up the hierarchy until an OR goal is found. In this case, control moves up through Borrow, which is an AND goal, and settles on the goal of Subtract-a-Column. (The AND/OR characteristic of goals is one of the details omitted from the presentation of the rules previously given.) The effect of this shifting of attention is to skip both the Decrement and Add10 operations of Borrow. In other words, instead of trying to decrement a zero, the procedure returns to examining the LFS column, which is still in its original form. Because Rule 4 has already run (and failed), Rule 2 is run, and Diff's precondition is violated. The four repairs generate these bugs:

Repair	*Bugs*
Skip	Blank-Instead-Of-Borrow-From-Zero (this bug had not been observed, yet)
Quit	Borrow-Won't-Recurse
Switch Arguments	Smaller-From-Larger-Instead-Of-Borrow-From-Zero
Dememoize	Zero-Instead-Of-Borrow-From-Zero

In short, with just three deletions and four repairs, we have generated nine different documented bugs and predicted the occurrence of two more.

Repair Rules, Both Usable and Unusable

It appears that good coverage of the bugs demands different repairs to the same precondition violation. How should this ability be incorporated into the theory? An obvious candidate is simply to list the repairs, leaving it to future research to determine why the list contains just the repairs that it does. Although pursuing accurate description before theorizing is a well-worn methodology, it defeats the immediate practical goal of being able to generate the bugs of a new procedure, such as multiplication. Hence, we have abstracted the repairs slightly so that they do not refer to the operations of subtraction alone. Thus, we have "switch arguments" rather than "use absolute difference for Diff." This makes what we have been calling repairs much more like general repair rules. That is, "switch arguments" can be thought of as a three step process:

1. Find two arguments of the operation whose data types are the same.
2. Switch them.
3. Install the modified operation as the response to the precondition violation.

The cost of using repair rules rather than prefabricated repairs is that the theory becomes slightly more complex. In particular, not all the repair rules apply to every precondition violation. Worse yet, some of the repair rules apply but do not cure the violation. In short, it is possible for a repair rule to be unusable. Some examples will help illustrate this complexity.

The repair rule "switch arguments" does not apply when Decrement's precondition is violated by a zero input. Because Decrement has only one argument, switching arguments makes no sense. Similarly, "switch arguments" doesn't apply to "WriteDigit ⟨place⟩ ⟨digit⟩." This primitive writes the given digit down on the paper in the given place. As the two inputs have different data types, "switch arguments" can't apply.

Diff has the precondition that both of its arguments are digits. However, in problems where the bottom row is shorter than the top row, certain deletions cause Diff to be called with a blank as the second argument. This violates the precondition. In this case, "switch arguments" applies, but executing the new version of Diff that it constructs causes another violation of the precondition. This is the second way a repair strategy can be unusable.

With the usable/unusable distinction in hand, one can make the following hypothesis:

> Any procedure that results from deleting a part of the correct algorithm and repairing each of the resulting precondition violations with a usable repair strategy will be an extant bug.

A "Set" of Repair Rules

There is a trade-off in designing a set of repair rules. Too few rules means an inability to generate some known bugs. But too many rules means predicting nonbugs. For example, there is a bug called Stutter-Subtract where the student reacts to nonrectangular problems by subtracting the leftmost digit in the bottom row from top digits that are over blanks. The Repair Theory analysis of this bug involves deleting Rule 3, which is repeated here:

> If the goal is to subtract a column, and the bottom of the column is blank, then write the top digit of the column as the answer.

This causes Diff to be entered with a blank as its second argument, which causes a precondition violation. To generate Stutter-Subtract, we need a repair rule, Refocus Right, which builds a repair for Diff that searches horizontally, moving rightward from the place where it expected to find its second argument. In short, Refocus Right seems necessary for generating Stutter-Subtract. However, Refocus Right can now be used to repair other precondition violations as well. Suppose, for example, that it is used to repair the zero precondition of

Decrement. This would generate a procedure that instead of borrowing across zero would decrement the column borrowed into. Moreover, if the top digit of the column borrowed into happened to be zero, the procedure would try to decrement the top digit of the column to its right. This bug has never been observed, and seems totally implausible. In short, one has a choice of failing to generate Stutter-Subtract, or generating nonbugs such as the one described.

The following set of repair rules seems fairly optimal. It generates 20 known bugs and a few plausible bugs that haven't been observed yet. We make no claim about this particular set, but are introducing it here, grouped under some suggestive headings, to illustrate another important trade-off:

Taxonomy of Repair Rules

I. Escape and Flee
   (a) Skip
   (b) Quit
   (c) Backup to last choice
II. Change the operation slightly
   (a) Switch Arguments of the same data type
   (b) Refocus Left
   (c) Refocus Down
III. Use an operation that worked in an analogous situation
   Increment for Decrement
IV. Dememoize

The headings are meant to suggest that the repair rules are really just instances of more powerful general purpose heuristics. Take the third category for example. The repair rule Increment for Decrement is just an instance of a general heuristic: If incrementing worked in an analogous situation, namely the "left half" of the regrouping operation of addition, then it ought to work here, in the "left half" of the regrouping operation of subtraction.

In one sense, it is quite heartening that the repair rules that fit best empirically can be viewed as instances of more general problem-solving heuristics. It is a little easier to believe that students bring a few powerful heuristics, perhaps developed elsewhere, to subtraction than to believe that they bring a diverse set of special purpose subtraction repair rules. However, if this view is taken seriously, then one must accept that the more powerful heuristics will apply all over subtraction, and this would lead to wild overgeneration (in particular, Refocus Right would have to be tolerated).

We have chosen to accept the intuitively more plausible position that repair rules are just special cases of general purpose problem-solving heuristics. To deal with the overgeneration, we propose to use a set of critics that test and filter out proposed repairs.

CRITICS

Critics signal that something about the current problem-solving state or solution is unusual. For instance, one might be changing the top digit in a column that has already been computed (the Refocus Right repair would, in general, trigger this Critic when it is trying to find a number to decrement after an attempt to decrement to zero has led to an impasse). The difference between a Critic and a precondition is that a Critic can always be ignored, whereas a precondition violation blocks the execution of its primitive. For example, the Critic mentioned in the foregoing could be ignored by Decrement, but under no circumstances could Decrement ignore the precondition that checks for a zero input.

Critics, most likely, are tacitly acquired by the student's observing and abstracting the patterns that all computations appear to satisfy—especially those that were produced by a teacher working through example solutions to homework problems. These abstractions fall naturally into several categories. The most obvious of these concern the form of what gets written (the answer and scratch marks indicating decrements). Some examples of Critics in this category are Critics Based on the Form of the Writing:

1. Don't leave a blank in the middle of the answer.
2. Don't have more than one digit per column in the answer.
3. Don't decrement a digit that is the result of a decrement.

Another category of Critics has to do with the information flow. They could also be induced from examples, or they may perhaps have been deduced from the more general beliefs about procedures. Some examples are Critics Based on Information Theoretic Properties:

1. Don't change a column after its answer is written (or more generally, each operation must make a difference to the answer).
2. Don't borrow twice for the same LFS column (or more generally, avoid infinite loops).
3. Both digits of the column should affect its answer (or more generally, there is no superfluous information in a subtraction problem).

Another kind of Critic could perhaps be induced from example, but is more likely a reflection of something the teacher said. Some examples are Critics Based on Semantics:

1. Unless borrowing is involved, a column's answer must not be greater than its top digit.
2. Subtraction is not commutative, so don't reverse the column.

It is clear that all students do not have all Critics (or if they do, they don't always heed them), as some known bugs violate obvious Critics (e.g., Blank-Instead-Of-Borrow). Because we cannot predict which Critics a student may have, the introduction of Critics could potentially reduce the theory's predictive power considerably. Nonetheless, they seem necessary for several reasons. One is, of course, to block overgeneration caused by using more powerful repair rules.

Another point in favor of incorporation of Critics is that they would fit very naturally into a process model for the theory. The process we have in mind is a simple generate-and-test problem solver for constructing the repairs. The generators are the repair rules. But even if the theory lacked Critics, there would have to be a tester for detecting unusable repairs. The "Critics" of the tester would be limited to the preconditions of the operations. If a precondition is violated, then the Critic would reject the proposed repair as unusable. By merely adding the above kind of Critics to the tester, overgeneration is avoided.

So far, preconditions have two roles in the theory, as triggers for repairs and as filters on repairs, whereas Critics are used only for filtering repairs. A particularly convincing demonstration of the reality of Critics would be the existence of bugs that require Critics to trigger the repairs that generate them. That is, we need a deletion that does not violate a precondition before it violates a Critic, and the repair of the Critic prevents the precondition violation, if any, from occurring. We have found only two such bugs so far and those are actually just subtle variants of other bugs. However, there are reasons for this paucity of bugs, reasons that are probably idiosyncratic to subtraction. When one examines the precondition violations generated by deletions, in most cases the violation occurs immediately after the deleted rule would have run. That is, there is simply no time between the deletion and the precondition violation to squeeze in a Critic violation. Hence, in most cases there is no opportunity for a Critic-triggered bug to be generated.

The two bugs whose generation involves Critics are subtle variants of two other bugs. The bug Don't-Decrement-Zero does borrowing across zero by changing the zero to a 10 instead of a nine. In problems where this zero is over another zero, the student is faced with writing 10 in the answer. Inasmuch as 10 is two digits long, there is a Critic violation. We have seen four different repairs to this Critic violation: (1) ignore it—just write the 10 in the answer; (2) write zero and carry one; (3) write the zero and ignore the one; and (4) write the one and ignore the zero. All these repairs have been observed in the addition bug data base!

The other bug that appears to involve a Critic is a subtle variation on Borrow-Across-Zero, which does borrowing across a zero without modifying the zero. Because the zero isn't changed to a nine, that column almost always requires a borrow that it would ordinarily not need. Consequently, the nonzero digit to its left gets decremented again (it was already decremented when the zero was borrowed across). Thus, in 307 − 19 = 198, the three is decremented twice, which violates a Critic. Usually, this Critic is ignored, but some students repair it. The only repair we have seen so far is to back up to the top-zero column, and do "switch arguments" to the resulting precondition violation in Diff. That is, when faced with a double-decrement situation, the bug retreats to the column causing the second borrow and writes its bottom digit down: 309 − 19 = 218. In short, both Critic bugs appear as subtle variants on standard bugs, but they do appear, and thus to a certain extent validate the incorporation of Critics into the theory.

CONCLUDING REMARKS

A theory of bugs has been outlined that embodies two important ideas. One is the methodologic idea of a generative theory: The theory succeeds to the extent that it can generate the observed bugs and, more importantly, that it cannot generate the absurd nonexistent procedures. The second idea that is fundamental to the theory is the notion of repairing impasses: When the student's knowledge is insufficient for solving a certain problem causing him or her to "get stuck," this impasse is repaired by the application of a small number of repair rules, selected so that the resulting procedure does not violate any Critics. Each such repair rule leads to a bug and all these bugs have the same origin, namely the given insufficiency in the knowledge base. In the theory sketched above, this insufficiency was defined by the absence of one of the rules that represents the correct procedure. The generative power of the theory comes, so to speak, from the multiplication of knowledge deficits (rule deletions) by repairs. This generative power can be illustrated with the results of an experimental computer implementation of the theory.

A rule system was constructed for correct subtract that allowed nine different rule deletions. Each deletion served as a formal representation of a knowledge or skill deficiency. For example, deleting the rule that tested for zero during borrow represented a lack of knowledge about borrowing across zero. These nine deletions caused nine distinct impasses. Application of repair rules to each impasse created 30 distinct procedures. Four of these were filtered out by the Critics. The remaining 26 procedures are predicted to be bugs. Of these 26 predicted bugs, 20 are well-documented bugs, and the other six have not been observed and hence are the theory's predictions for future bug discoveries. To give a sense of context, it is worth pointing out that when the implementation mentioned in the foregoing was first tested in September 1979, only 17 of its 26 procedures were known bugs. The present figures are from December 1979. So, in the intervening 3 months, three of the predicted bugs were actually found. We fully expect to find the other six someday. The 20 bugs that have been found are listed in Table 9.2.

TABLE 9.2
The 20 Bugs Generated by an Experimental Implementation of the Theory

Can't-Subtract
Quit-When-Bottom-Blank
Stutter-Subtract
Blank-Instead-Of-Borrow
Doesn't-Borrow
Smaller-From-Larger
Zero-Instead-Of-Borrow
Borrow-No-Decrement
Smaller-From-Larger-With-Borrow
Zero-After-Borrow
Stops-Borrow-At-Zero
Borrow-Won't-Recurse
Smaller-From-Larger-Instead-Of-Borrow-From-Zero
Zero-Instead-Of-Borrow-From-Zero
Borrow-From-Bottom-Instead-Of-Zero
Borrow-Add-Decrement-Instead-Of-Zero
Don't-Decrement-Zero
Borrow-From-Zero-Is-Ten
Borrow-Across-Zero
Borrow-From-Zero

In short, with just nine knowledge deficiencies, 20 observed bugs have been accounted for. This illustrates the generative power of the theory.

However, much work remains to be done. There are currently 88 bugs in our data base. Only 20 of these were generated by our naive "Repair Theory." To generate the remaining bugs, other kinds of knowledge deficiencies than the ones represented by rule deletions need to be formalized. Deletion of rules from the correct procedure is only one way to generate knowledge states. Other techniques are being investigated.

As an example, mechanisms for generalization or learning by example that were pioneered by Winston (1970) can be used to generate more knowledge states that can be repaired to form bugs. Imagine that a student has only practiced borrowing on two-column problems. This could lead him or her to have an overly specific idea of where to perform the decrement. In particular, a formal learning program could produce a knowledge state wherein the procedure borrows from the column that is both the leftmost column and the column adjacent to the one being borrowed into. Given this knowledge defect, repair theory makes a number of bug predictions. When the overly specific procedure is executed on a three-column problem with a borrow on the units column, an impasse occurs because there is no column that is both the leftmost and the left-adjacent column. Application of the No-op repair rule leads to the bug Borrow-Don't-Decrement-Except-Last. Other repairs predict different bugs. For example, dropping the adjacency constraint generates the bug Always-Borrow-Left. In short, when other formalisms are employed to augment rule deletion, many more of the known bugs can be generated formally.

Note that such extensions stay within the basic framework of impasses, repairs, and Critics. This framework, operating within the methodology of a generative theory, shows great promise for explaining the existence of a wide variety of systematic errors in procedural skills.

Repair Theory may even have explanations for more than just the occurrence/nonoccurrence of bugs. It has the potential for making predictions about bug stability. As noted earlier, there is anecdotal evidence that bugs can appear and disappear in a matter of days. Students have been observed who had a bug one day and a different bug a few days later. Repair Theory provides an account of this "bug migration" phenomenon. Suppose that on the first occurrence of an impasse on a test, the student created a repair and used it throughout the test by storing it in some kind of short-term, temporary memory. Suppose further that when he or she takes the second test, a few days later, he or she has forgotten the repair made previously. Hence, when an impasse is encountered, he or she may use a different repair and hence exhibit a different bug. This is the theory's explanation of bug migration. However, it also makes a prediction. It predicts that bugs that migrate into each other will be related in that they are different repairs to the same impasse. For example, Borrow-Across-Zero would migrate into Stops-Borrow-At-Zero, but not into Borrow-From-Zero. We have anecdotal evidence for this kind of migration and have designed experiments to verify it.

In sum, the theory not only provides a framework and some preliminary results that explain the existence of systematic errors, but has ramifications for an account of instabilities of bugs and perhaps other phenomena as well.

REFERENCES

Brown, J. S., & Burton, R. B. Diagnostic models for procedural bugs in basic mathematical skills. *Cognitive Science,* 1978, *2,* 155–192.

Burton, R. B. DEBUGGY: Diagnosis of errors in basic mathematical skills. In D. H. Sleeman & J. S. Brown (Eds.), *Intelligent tutoring systems.* London: Academic Press, 1981.

Hetzel, J. Private communication, 1979.

VanLehn, K. On the representation of procedures in Repair Theory. Technical Report, University of Pittsburgh, Learning Research and Development Center, 1980.

Winston, P. H. Learning structural descriptions from examples. Artificial Intelligence Laboratory Technical Report, AI-TR-231, Massachusetts Institute of Technology, 1970.

10 Syntax and Semantics in Learning to Subtract

Lauren B. Resnick
University of Pittsburgh

This chapter is concerned with the role of meaning and understanding in the acquisition of computational skill. Until quite recently, discussions of meaningfulness in arithmetic learning have been characterized by a confrontation between those who advocate learning algorithms and those who argue for learning basic concepts. This confrontation, however, is neither necessary nor fruitful. In fact, it serves to direct attention away from the ways in which understanding and procedural skill may mutually support and influence one another.

Proponents of conceptually oriented instruction have frequently argued that a major cause of difficulty in learning arithmetic is a failure to relate rules of computational procedure to their underlying mathematical concepts. Substantial efforts on the part of mathematics educators have resulted in instruction intended to display for children the structure of important concepts such as the base system and positional notation, on the assumption that procedural algorithms would be easily acquired and retained when the conceptual basis was obvious. The data presented here show that intuitions concerning the importance of conceptual understanding are correct in a crucial respect: difficulties in learning are often a result of failure to understand the concepts on which procedures are based. But the data also show that even when the basic concepts are quite well understood, they may remain unrelated to computational procedure. Thus the conceptual teaching methods of the past were inadequate to the extent that they taught concepts *instead of* procedures and left it entirely to students to discover how computational procedures could be derived from the basic structure of the number and numeration system. Our research thus poses an important new problem for mathematics instruction: devising methods that help students to explicitly link meaning and procedure.

DISTINGUISHING SYNTAX AND SEMANTICS

Written subtraction can be analyzed as an algorithm defined by a set of *syntactic* rules that prescribe how problems should be written, an order in which certain operations must be performed, and which kinds of symbols belong in which positions. Although the syntax may reflect an underlying *semantics,* or meaning, an algorithm need not include any explicit reference to the semantics in order to be successfully performed. Figure 10.1 shows an algorithm for subtraction that is

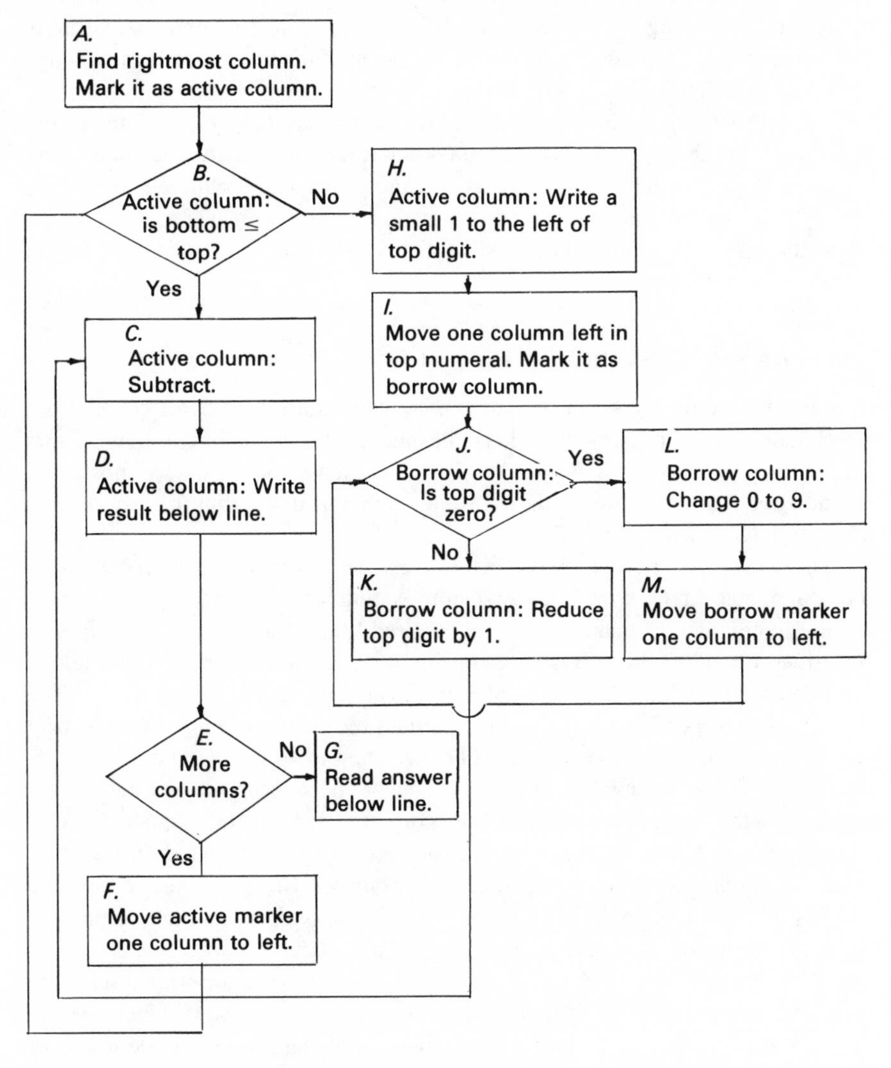

FIG. 10.1. Syntactic algorithm for subtraction.

entirely syntactic in nature. Followed strictly, it can solve any subtraction problem written out in aligned column format. Some of the syntactic rules are specified in the figure (e.g., writing 1 in a particular position at *H* and reducing by one at *K*). Other syntactic rules the algorithm obeys are not specified in the figure. For example, there can be only one digit per column, and each column must be acted upon at least once.

Because performing this algorithm requires no understanding of the base system, I claim that it includes no semantics. That is, the system performing this algorithm does not need to know such things as the fact that the small 1 inserted as part of borrowing really represents 10 (when it is in the rightmost, or units column), or that borrowing involves an exchange of quantities between columns that does not change the total quantity represented. Even the reason for borrowing is omitted. This kind of semantic knowledge would justify the syntax of the algorithm, but the algorithm can be run without reference to the semantics. The evidence I present here suggests quite strongly that many children may learn the syntactic constraints of written subtraction without connecting them to the semantic information that underlies the algorithm. This can lead to systematic errors in performance.

Evidence for the Syntactic Nature of Subtraction Errors

Initial evidence for the syntactic nature of written subtraction errors comes from the extensive work of Brown and his colleagues on the nature of children's errors in subtraction (Brown & Burton, 1978). They have shown that a substantial portion of children's errors result not from random mistakes, but from *systematically* following wrong procedures.

The wrong procedures are variants of correct ones; they are analogous to computer algorithms with bugs in them. A finite number of bugs, which in various combinations make up several hundred "buggy algorithms," have been identified for subtraction. Bugs presumably arise when a subtraction problem is encountered for which the child's algorithm is incomplete or inappropriate. The child tries to respond anyway and either applies the incomplete algorithm leaving out necessary steps, or tries to repair the algorithm to adjust to the new subtraction task. The resulting bug respects some of the constraints—syntactic and semantic—that are embedded in a full, correct algorithm, but violates others. (See Brown & VanLehn, this volume, for details of this "repair" theory.)

By examining some common subtraction bugs and considering the possible origins of each, it is possible to decide whether the bugs represent violations of primarily semantic or syntactic constraints. This analysis, part of which is presented in the following section, suggests that most buggy algorithms respect the syntactic requirements of written subtraction, whereas the constraints that are relaxed are the ones that express the semantics of the base-ten system and of subtraction. This informal analysis appears consonant with both Brown and VanLehn's theory and with another theory by Young and O'Shea (1981). In the latter theory children construct buggy algorithms because they either have never learned the complete standard algorithm or have forgotten parts of it.

Figure 10.2 names a number of common bugs in order of their frequency, describes them, and gives examples of each. In what follows, I consider possible sources of these bugs and characterize each bug as semantic or syntactic in character.

1. Smaller-From-Larger. Repair theory suggests that this very common bug results from "switching arguments" to respond to a situation in which the system cannot make its normal move of subtracting the bottom from the top number in a column. In other words, the system makes the test at *B* in Fig. 10.1, but doesn't know how to borrow and decides that the subtraction should be done in the opposite direction. Young and O'Shea's analysis suggests that this bug derives from simply not making the test and is the normal or default way for the system to proceed *unless* the test is made and the various borrowing rules are thereby evoked. In both of these interpretations all the syntax of written subtraction without borrowing is respected. What is violated is the constraint that the bottom quantity *as a whole,* be subtracted from the top quantity *as a whole.* The semantics of multi-digit subtraction includes the constraint that the columns, although handled one at a time, cannot be treated as if they were a string of unrelated single-digit subtraction problems.

2. Borrow-From-Zero. Both repair theory and the Young and O'Shea analysis suggest that this bug derives from forgetting the part of the written procedure that is equivalent to steps *M–J–K* in Fig. 10.1 (moving the borrow marker left, and reducing the new column). The bug respects the syntactic requirement that, in a borrow, there must be a crossed-out and rewritten numeral to the left of the active column. It also respects the syntax of the special case of zero, where the rewritten number is always 9. However, it ignores the fact that the 9 really results from borrowing one column further left (the hundreds column) moving 100 as 10 tens into the tens column, and *then* borrowing from the 10 tens leaving 9 tens, or 90 (written as 9).

3. Borrow-Across-Zero. Repair theory offers two different derivations of this bug. The first is that this bug arises from the child's search for a place to do the decrementing operation with the condition that the column not have a zero in the top number. This would happen when the child doesn't know how to handle zeros or thinks they have "no value" and thus can be skipped. This solution respects the syntactic constraint that a small 1 must be written in the active column and that some other (nonzero) column must then be decremented. But the

1. **Smaller-From-Larger.** The student subtracts the smaller digit in a column from the larger digit regardless of which one is on top.

   326 − 117 = 211 542 − 389 = 247

2. **Borrow-From-Zero.** When borrowing from a column whose top digit is 0, the student writes 9 but does not continue borrowing from the column to the left of the 0.

   602 − 437 = 265 802 − 396 = 506

3. **Borrow-Across-Zero.** When the student needs to borrow from a column whose top digit is 0, he skips that column and borrows from the next one. (This bug requires a special "rule" for subtracting from 0: either 0 − N = N or 0 − N = 0.)

   602 − 327 = 225 804 − 456 = 308

4. **Stop-Borrow-At-Zero.** The student fails to decrement 0, although he adds 10 correctly to the top digit of the active column. (This bug must be combined with either 0 − N = N or 0 − N = 0.)

   703 − 678 = 175 604 − 387 = 307

5. **Don't-Decrement-Zero.** When borrowing from a column in which the top digit is 0, the student rewrites the 0 as 10 but does not change the 10 to 9 when incrementing the active column.

   702 − 368 = 344 205 − 9 = 1106

6. **Zero-Instead-Of-Borrow.** The student writes 0 as the answer in any column in which the bottom digit is larger than the top.

   326 − 117 = 210 542 − 389 = 200

7. **Borrow-From-Bottom-Instead-Of-Zero.** If the top digit in the column being borrowed from is 0, the student borrows from the bottom digit instead. (This bug must be combined with either 0 − N = N or 0 − N = 0.)

   702 − 368 = 454 508 − 489 = 109

FIG. 10.2. Descriptions and examples of common subtraction bugs. (From "Diagnostic models for procedural bugs in basic mathematics skills," By J. S. Brown and R. R. Burton, *Cognitive Science,* 1978, *2,* 155–192 and from personal communication with J. S. Brown, R. R. Burton, and K. VanLehn.)

semantic knowledge that the increment and decrement are actually addition and subtraction of 10 is ignored (or not known). Repair theory's second derivation, which agrees with Young and O'Shea's analysis, produces this bug by simply deleting the rule that changes 0 to 9 (*L* in Fig. 10.1). This too is a completely syntactic derivation, for it allows deletion of a rule without reference to the semantic information that justifies the operation.

4. *Stop-Borrow-At-Zero*. Both repair theory and Young and O'Shea's analysis interpret this bug as simply omitting a rule or an operation. Steps *I–J–K* of Fig. 10.1 are simply skipped. This bug fails to obey both syntactic and semantic constraints. Syntactically it produces only the increment part of the borrow operation—the 1 in the active column—but does not show a crossed-out number or the change of a 0 to a 9. Semantically it violates the justification for the borrow increment—that is, in order to add a quantity to the active column an equivalent quantity must be subtracted from another column.

5. *Don't-Decrement-Zero*. The change of 0 to 10 in this bug is the proper "semantic" move after borrowing from the hundreds column. But it produces an outcome that the child may not have encountered and thus does not respond to appropriately. Failure to change the 10 to 9 may result from a syntactic constraint that each column be operated on only once. This syntactic constraint is not "correct," but might be reasonably inferred from extensive experience with problems that contain no zeros. If so, the syntactic constraint is in direct opposition to the semantic demands of the situation.

6. *Zero-Instead-Of-Borrow*. Like Smaller-From-Larger, this bug simply avoids the borrowing operation altogether, while observing all of the important syntactic constraints of operating within columns, writing only one small digit per column, and the like. This bug, however, does not violate the semantics of the digit structure as blatantly as the Smaller-From-Larger bug. In fact, a child producing this bug may be following a semantics of subtraction that generally precedes any understanding of negative numbers. In this inferred semantics of subtraction, when a larger number must be taken from a smaller, the decrementing is begun and continued until there are no more left—yielding zero as the answer.

7. *Borrow-From-Bottom-Instead-Of-Zero*. This bug seems purely syntactic in the sense that the search for something to decrement seems to lead the child to ignore the digit structure and the semantics of exchange that justifies borrowing within the top number. But it does produce a "funny-looking" solution, so it would probably be generated only by a child whose syntactic rules did not specifically require that all increments and decrements be in the top number.

A CLOSER LOOK AT CHILDREN'S SEMANTIC AND SYNTACTIC KNOWLEDGE

If bugs result from weak application of semantic constraints, a first hypothesis is that children are simply unfamiliar with the semantics of the base-ten system. Data we have collected permit us to examine children's semantic and syntactic knowledge quite directly. These data lead me to conclude that this simple hypothesis is false—that instead, children are likely to know a good deal about the base system but still be unable to use their semantic knowledge to support written procedures for arithmetic.

The data are from four children who were followed between November and May of the school year in which they first learned addition and subtraction with regrouping. The children were all in a nongraded, individualized mathematics curriculum at the beginning of the study. Three were second-graders and one was a third-grader. In November these children had recently passed the criterion test for a unit in which they were required to read and write numerals up to 100 and to interpret these as compositions of tens and ones. We reinterviewed the children in February, after each had encountered instruction in addition and subtraction with regrouping but had not yet passed the criterion test for those skills, and again in May, at which time all had passed the curriculum test for addition and subtraction with regrouping.

All interview sessions were individually administered and semistructured, with a planned sequence of problems presented in a standard format. Probes by the experimenter and some attempts to explain or even demonstrate a procedure were permitted. Learning obviously took place in the course of the interviews—sometimes in response to the experimenter's "teaching," sometimes as a result of inventions by the children. We have used the speed and character of these learning incidents to infer the children's knowledge base.

The content varied over the three interview periods. In November, we focused on the children's understanding of the semantics of the base system. Most of the tasks required each child to represent written numerals (10 through 99) in concrete forms and to add and subtract in the concrete representations. Dienes blocks, color-coded chips, bundles of sticks, or pennies and dimes were used for the representations. In February, virtually all the tasks were addition and subtraction problems, presented in both written and concrete form. We paid particular attention to the extent to which the child made correspondences ("mappings") between written and concrete representations of these processes. The May interviews replicated those of February but added a special session in which the child was asked to teach a hand puppet to add and subtract. This allowed the children to give explanations and justifications for their addition and subtraction procedures in a less self-conscious way than by simply explaining to an adult why certain routines were used. The results of these interviews are summarized in the following section.

Semantics of Concrete Representations of the Base System

Three of the four children—Amanda, Alan, and Anton—demonstrated immediate and strong knowledge of the base system in concrete representations. They showed this by:

1. Immediately using the designated color chip, bundle, or block shape to represent tens.
2. Counting by tens and then ones as they "read" concrete displays or counted out the objects to construct those displays.
3. When comparing concrete displays, sometimes counting only the tens—thus relying on a canonical (i.e., no more than nine items of any one denomination) display in which the representation of the higher number will necessarily have more tens.
4. When counting large numbers of units, grouping them into piles of ten, in order to count by tens.
5. Recognizing the conventionality of the codes, as when Amanda said that the bundles of sticks might have nine or eleven sticks each, but: "Let's say they all have ten."

One child, Alan, apparently learned the power code in the course of the initial interviews. He began by counting all blocks, chips, etc., as units, but switched to counting by tens when the experimenter asked him to represent a large number and provided too few blocks to represent it in units. His speed in picking up our conventions suggests that it was only the conventions that he had to learn, not the base-ten semantics that they represented. By contrast the fourth child, Sandra, seemed to be truly acquiring the semantic knowledge of the base system over the course of the period in which we studied her. In November, although the experimenter's prompts would lead her to count by tens and use this code for a few problems, she would revert to counting all denominations as ones whenever new representations were presented or the experimenter did not remind her of the convention. With the Dienes blocks she used a "compromise" solution, in which each of the individual squares on the ten bar was counted. She thus respected the conventional coding, but did not really benefit from its "ten-ness." Even in May (although by this time she always used the code representations), Sandra still counted the individual squares on the ten bar. Of the children in our group, Sandra clearly had the weakest command of the base-system semantics and the least tendency to use it in constructing shortcuts.

Procedures for Adding and Subtracting With Concrete Representations

In the November interviews the only addition and subtraction problems given to the children were in the context of a store game using the penny-and-dime representation. This was the representation with which all four children had seemed most comfortable. All the children were in command of the basic semantics of addition and subtraction. That is, they knew how to add by combining and then recounting objects that originally represented two numbers, and they knew how to subtract by taking away a specified quantity from a given representation of number. All could do this without error for two-digit representations, as long as there was no need for regrouping.

Addition. The equivalent of the carry in written addition is trading ten pennies for a dime in the money representation. In November, none of the children ever initiated a trade of pennies for dimes. However, when the experimenter said, "The store won't accept so many pennies," each child traded ten pennies for a dime without further prompts. Thus, although the children knew that exchanges were possible, they did not naturally tend to use the concrete representations in a manner that corresponded well to the rules of the written algorithm.

Further evidence from the protocols suggests that a preference for canonical form—which would match the written mode—appeared rather late in the development of the children's understanding of the base system. For example, Amanda, whose mental arithmetic performance and general facility with the various concrete representation tasks suggested a very early and strong command of the base system, seemed to have the *least* preference for canonical displays in November. She instead seemed to be experimenting with the various ways in which a given number could be represented. She constructed noncanonical displays on various problems and then converted them to canonical, or vice versa. In general, she made far more trades and exchanges (always ten-for-one or one-for-ten, in keeping with her strong command of the ten-ness in the base system) than any of the other children. By February, however, Amanda had begun to prefer canonical form. She initiated trades in adding and tended to do her trading sequentially—that is, each time that she accumulated ten blocks in a long addition problem, she traded for the next block size.

Sandra, the child who was still counting by ones in May and not using the ten-ness of the system, did not at any time initiate trades to canonical form on her own. But she did show a response to the experimenter's rule of "no more than nine per column" that suggests that she had adopted the rule—without reference to its rationale—as an arbitrary constraint to be followed at all costs. The following protocol segment illustrates this:

14 + 9 + 33. S: (Sets out one tens block and four units blocks; nine units blocks; three tens blocks and three units blocks.) "Ten, twenty, thirty, forty, forty-one, . . . fifty-one, . . . fifty-six." *E:* "Now what if I told you there could not be more than ten in a column? How could you get rid of them?" *S:* (Hesitates.) *E:* "Could you trade? You could trade ten ones for a ten." *S:* (Trades, putting the ten in the tens column.) *E:* "Do you have the same as before?" *S:* (Counts by tens, continuing with the ones.) "Fifty-six." *E:* "Still the same?" *S:* "Yes."

75 − 49. *S:* (Sets out seven tens and five units.) "I don't have nine." *E:* "Can you trade one of these?" *S:* "But then I would have ten" (i.e., more units than allowed in a column).

Sandra accepted the experimenter's rule for addition and then applied it to subtraction as well, showing resistance to the suggestion that she trade to solve a subtraction problem.

This performance suggests why *non*preference for canonicity may be an important developmental stage. Perhaps a strong rule specifying canonicity can interfere with learning about situations other than addition until children have come to understand the situations in which canonicity should be preferred.

Subtraction. The equivalent of borrowing in concrete representations is the process of getting more units by trading a ten for ten units. The problem initially posed to the children in November was to take 61 cents (which all of the children did by taking six dimes and one penny) and then give the experimenter 37 cents. There are two critical aspects of this performance. The first is whether the child *initiates* a trade-down or needs to be prompted. The second is whether the trade is "fair"—i.e., whether the child always makes a one-for-ten trade.

In November, only Amanda (who had the most highly developed understanding of the base system) both initiated the trade and made a one-for-ten trade with no prompting or explanation. Sandra, the weakest of the four children, needed to be explicitly told to trade on almost every trial. When unfair trades were made they almost always followed a pattern of trading the dime for only as many pennies as were needed to give the experimenter the number she requested. For example, in the 61 − 37 problem Alan traded a dime for six pennies which, together with the penny he already had, allowed him to give seven pennies to the experimenter.

The children improved between November and May, so that by May all but Sandra were initiating trades and making only fair trades. Even Sandra needed only an initial and weak ("get more") prompt. Because there is little practice in the children's school curriculum on subtraction using these concrete representations, it seems reasonable to suppose that the much more skillful and semantically correct performance they all showed by May was not the result of practice on the subtraction routine itself, but of the development of general semantic knowledge of the base system and subtraction. This suggests that the later trades are driven by the need to get more units (or tens) and that the requirements of equivalent (one-for-ten) trades are well internalized.

Semantics of the Writing Code

All the children could interpret written numbers as "x tens and y ones." Performance on other tasks that tapped knowledge of the writing code, however, suggests that this may not reflect a very rich knowledge of the semantics of written numerals. We observed the following:

1. Three of the children correctly represented two-digit numerals with concrete representations on the first trial or after only a few prompts by the experimenter. They could also write the numerals when shown a display in one of these concrete representations. This shows some knowledge of the writing code. However, this knowledge appeared to be only weakly linked to the concrete representations. It was common for the children to count all the chips or blocks whether in canonical form or not, say the numeral aloud, and then write this numeral without matching its digits to the concrete display.
2. Three of the four children were able to use expanded notation cards to construct numerals. For example, 98 was constructed from 90 and 8, with the 8 placed on top of the 0. There is some evidence, however, that the solutions to these expanded notation problems were more syntactic than semantic, as the children seemed to be trying different ways of putting the cards together until they found something that looked right. All the children demonstrated some difficulty with zeros (as in the number 708) when using these cards.

Two of the children (including the one who did not use expanded notation cards correctly) gave us spontaneous evidence of a deeper understanding of the written code. This came in the form of:

3. Occasional comments when working on various problems. For example, Alan, when comparing the numerals 9 and 90, said the 90 was larger because the 9 "doesn't even have ten."
4. Solving written problems mentally and then writing down the answers. Amanda used a mental arithmetic strategy in which she partitioned two-digit numbers into tens and ones and then operated separately on the two sets of values. For example, she solved 37 + 25 as follows: "Thirty plus twenty is fifty. Fifty-seven. Fifty-seven, fifty-eight, fifty-nine, sixty, sixty-one, sixty-two." Then she wrote 62, aligning the digits in the proper columns.

Procedures for Written Subtraction and Addition

Written addition and subtraction problems were first presented to the children in February and were repeated in May. Except for a few special probes, only two-digit problems in subtraction were used, so we were unable to observe any of the zero bugs described earlier. From these protocols, however, we can chart the development of three key components of written algorithms: right-to-left rules of procedure, the carry procedure, and the borrow procedure. Three of the children—Sandra, Anton, and Alan—can be followed with respect to these components. Amanda's use of mental arithmetic allowed her to avoid the written algorithms altogether.

Right-to-Left Direction. The three children all showed a rather slow and hesitant development of the right-to-left rules of procedure. Each had learned in school that right to left is the "correct" way to do the problems. When asked why they should start at the right they typically answered, "So you get the right answer" or "It's easier," but never, "To make borrowing or carrying easier" or a similar comment. In February each child began by working left to right and switched to a right-to-left direction after encountering difficulty. These switches were spontaneous; the experimenter never suggested working in the other direction. They were not, however, full-fledged inventions, as it is quite clear from the protocols that a procedural rule learned earlier was being invoked to cope with difficulties encountered.

Carry Procedure. Sandra, the weakest of the three in her command of the semantics of the base system as represented in concrete materials, performed the carry procedure perfectly in February. (It had been taught to her in school by then.) Anton initially worked left to right, but quickly corrected himself. In February, Alan almost completely avoided carrying by counting on his fingers and then writing the answers, and he continued to show some difficulty with carrying in May. Amanda did not use the written carrying algorithm at all, even in May. Thus, strong semantic understanding (as shown by manipulations of the concrete representations and use of mental arithmetic) in no way guarantees ability to use the carry procedure in written addition.

Borrow Procedure. As might be expected, this procedure was harder to learn than carrying. Alan, Anton, and Sandra in the February and May interviews each showed the Smaller-From-Larger bug, but each then spontaneously switched to a borrow operation. By May, Sandra and Anton were using the correct, school-taught written algorithm, notating both increments and decrements. Alan repeatedly stated that he was supposed to borrow, but even in May he had some difficulty remembering the exact procedure. Amanda first performed in the borrow procedure in May, with many false starts and bugs. Again, the two children with the strongest knowledge of base system semantics—Alan and Amanda—seemed to have the weakest control of the written algorithm.

Mapping

The final question to be addressed on the basis of these data is the extent to which the children had connected their knowledge about the base system (as displayed in their work with the concrete materials) and their knowledge about the written procedures for subtraction and addition. A general response can be made by simply noting again the lack of correlation between knowledge of the base-system semantics and knowledge of the written algorithms.

More specifically, our interview data permit a closer examination of three levels of mapping between the written algorithms and the concrete materials: (1) code mapping—the extent to which the child recognizes that the shape or color of the concrete materials codes the same information as position (column) in the written numerals; (2) result mapping—the extent to which the child expects procedures in the written system to yield the same answers as procedures in the concrete materials; and (3) operations mapping—the extent to which the child can identify equivalent operations in the written and the concrete systems and can model a written algorithm in a concrete mode.

Code mapping was at least weakly present in all the children. This was shown by their quickly acquired ability to represent written numerals in concrete forms and to write numerals that corresponded to the concrete representations. However, none of the children insisted on an exact correspondence between the written and the concrete form.

There were distinct differences among the four children with respect to result mapping. Although we did not systematically probe for this, Amanda and Alan gave clear evidence of expecting blocks and writing to yield the same answers. During the May interview, for example, Alan corrected his written subtraction after doing the same problem with blocks. He trusted his blocks answer, and expected writing and blocks to yield the same answer, showing this by doing a new or amended written procedure. Anton, by contrast, seemed undisturbed by getting different answers in blocks than in written subtraction.

Although Alan expected the same answers from blocks and writing, he did not use detailed operations mapping to correct his written algorithm. Instead, he seemed to search his memory for a previously learned procedure that would correct his errors. There was also other evidence that these children were not doing operations mapping. With the possible exception of Amanda in May, at no time did we see the children examining the steps in a blocks procedure as a way of helping themselves perform a written algorithm.

LINKING SYNTAX AND SEMANTICS

If the preceding analysis is correct, then children's difficulty with place value in addition and subtraction does not necessarily derive from an *absence* of semantic knowledge about the base system. Rather, it results from an inadequate *linking* of the semantics of the base system with the syntax of the written algorithms. This suggests that for many children who have difficulty learning or remembering the rules for written arithmetic, instruction that explicitly links semantic and syntactic knowledge may be useful. In addition, initial instruction that stresses the semantic properties of the addition and subtraction algorithms should help block the difficulties and buggy routines that might arise later.

We have pilot tested with three other children a remedial teaching procedure that explicitly forces a mapping at the operational level between block subtraction and written subtraction. The procedure requires that the child perform the same problem using Dienes blocks and writing, alternating between the two representations. The writing thus serves as a record of the blocks actions. Conversely, the blocks justify the steps in the written algorithm. Figure 10.3 depicts the alternating blocks and written steps for a simple problem. We demonstrated the procedure to the children along with a great deal of verbalization to explain why each step was taken. Following the demonstration there were repeated trials in which the children did more and more of the work themselves. During this process more complex problems, including three-digit problems, were intro-

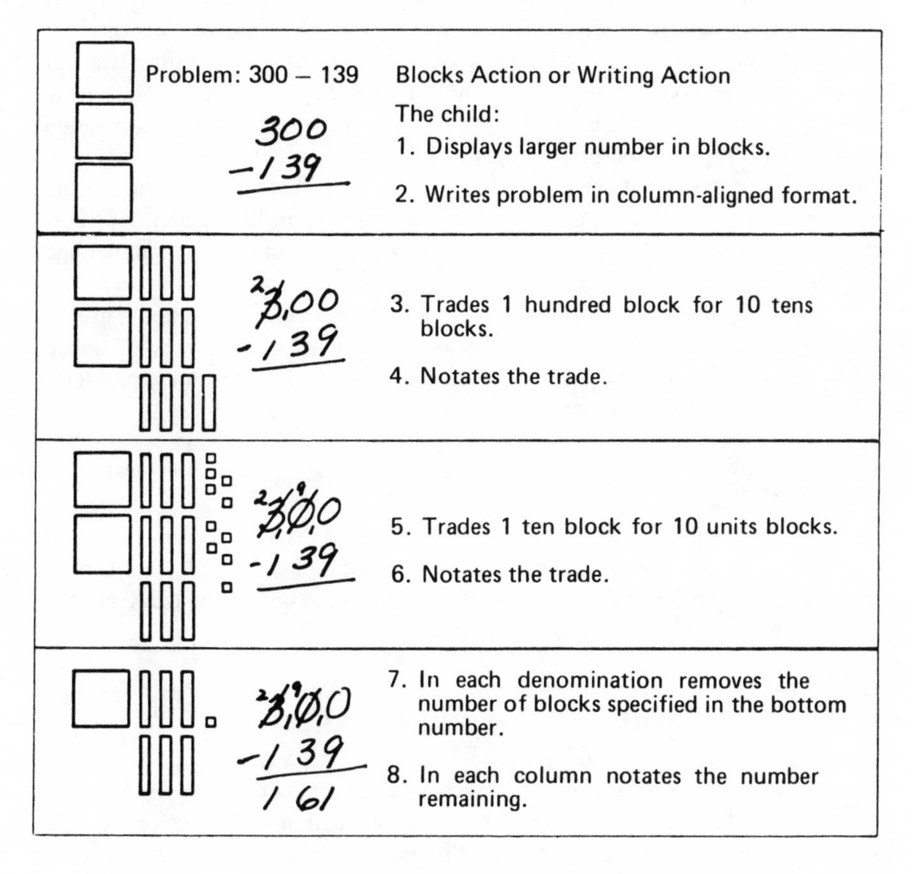

FIG. 10.3. Solution of a subtraction problem with blocks and a written algorithm.

duced. Eventually the child took over the entire process of manipulating blocks and writing the corresponding algorithmic step. Typically, within about forty minutes each student was performing the task smoothly. At this point the blocks could be removed (sometimes after a brief intermediate period in which "imaginary" blocks were manipulated) and purely written work undertaken.

The three children who have been taught this mapping procedure each began with a different diagnosed bug. Laura was doing Smaller-From-Larger; Molly, Don't-Decrement-Zero; and Ann, Borrow-Across-Zero. Each was brought back for a follow-up interview three to six weeks after the instruction. On this occasion each performed a series of written problems with ease. The bug was gone and no new ones had arisen. Further, each child remembered clearly what her original bug was and could explain what she had learned about why the bug was wrong.

All three children were questioned about what the various digits stood for, why they were being crossed out and written in, etc. Their answers suggest that they understood *why the algorithm works,* especially as their wording was not the same as the wording the experimenter used during instruction. They were also able to show correctly which blocks stood for the special borrowing notation digits (for example, a ten block for the small 1 written before the units digit), indicating a mapping between the two representations that had not been present prior to instruction. Finally, each child showed some interesting pattern of transferring semantic knowledge by inventing a procedure or an explanation that had not been taught.

Molly's performance was particularly striking, because she invented a justification for a (correct) written notation that did not exactly match the one produced during mapping instruction. The experimenter had asked her to do the problem, 2003 − 1467, and she had written:

```
 1 9 9
 2̸ 0̸ 0̸ 13
-1  4  6  7
-----------
```

Under questioning, she said she had borrowed one thousand; we then asked her to tell us where that thousand had been placed. She replied, "Well, 100 is right here (pointing to the 1 of the 13 and to the 9 in the tens column), and 900 is right here (pointing to the 9 in the hundreds column)."

WHY MAPPING WORKS

There is, of course, considerably more work to be done to establish the range of conditions under which this kind of mapping instruction will be effective. We will want to explore what kinds of bugs this instruction is capable of eliminating, and what initial knowledge of place value and addition and subtraction semantics is required for mapping to be effective. We will also need to examine the possibility that certain bugs inherent in the blocks routines may unwittingly be adopted into the writing algorithm along with all the semantic corrections. Finally, we will want to consider the possibility that other representations—sticks, abaci, expanded notation, for example—may be equally or more powerful than Dienes blocks for the purpose discussed here, and the strong possibility that mappings between three representations may be even more powerful than between two in producing understanding of the written algorithms.

There is, however, enough evidence now in hand about the power of mapping instruction to warrant building a systematic theory of how mapping works to build understanding and to correct procedural errors. Although I do not attempt a formal theory here, in the remainder of this chapter I consider the possible shape of such explanatory theories. In the process, I review the kind of empirical evidence that might sharpen the theory-building effort.

There seem to be two possible accounts for the effects of mapping. One, the *prohibition* explanation, focuses on how pairing of each step in the written algorithm with a parallel step in blocks might serve simply to prohibit wrong operations in the writing. A second possibility, the *enrichment* explanation, is that semantic knowledge initially embedded in the blocks algorithm is, by mapping, applied to the rules for writing so that the newly enriched knowledge structure then eliminates bugs.

Prohibition

It is possible that most of the effect of the mapping instruction derives not from acquiring a deep understanding of the semantics of subtraction, but simply from the external constraints imposed by rules of the instructional situation. If the subtraction with blocks is performed correctly and if the rule of alternating operations in the blocks and the written representation is followed exactly, most of the known written subtraction bugs are impossible because they cannot be modeled with blocks without violating basic exchange principles. The way in which mapping between blocks and writing could prohibit writing bugs can best be appreciated by considering individual bugs.

Smaller-From-Larger is prohibited because in the blocks subtraction routine only the top number is represented in the blocks. The task is to remove the number of blocks specified in the bottom digits. As long as all digits of the top number are represented in blocks before any subtraction operations begin, there is no way of reversing top and bottom digits in a particular column, because only the top digit has been displayed.

The various bugs that arise when it is necessary to borrow from a column with zero as its top number are also prohibited in the mapping situation, provided the child knows that a fair trade must be made. When working with blocks the organizing goal is to get enough blocks to permit removing the specified number. These blocks must always come from the top number as it is the only one represented. This constraint forces the child to go left until blocks are found. The child cannot add blocks to the zero column (Borrow-From-Zero) until something has been "borrowed" from the hundreds column. Similarly, Stops-Borrow-At-Zero is prohibited because the child cannot add blocks to a column without having borrowed from another column. Borrow-From-Bottom-Instead-Of-Zero is prohibited because the bottom number is not even represented. Borrow-Across-Zero is prohibited by details of the exchange rules: having borrowed from the hundreds column, the normal exchange (which we assume the child knows) will be for ten tens. Because the blocks have been arranged in columns for mapping instruction, the new blocks must be placed in the tens column, and this in turn requires some notation in that column. Also, a second trade is needed to fulfill the goal of getting more blocks in the active column. This second trade would have the effect of prohibiting Don't-Decrement-Zero, which changes the 0 to 10 but does not then decrement it to 9.

The point of this analysis is that it demonstrates that, to the extent that the rules of block exchanges and block subtraction are followed along with the rules of mapping, most of the operations associated with buggy algorithms are simply not possible.[1] Perhaps, then, the power of mapping instruction lies largely in providing a high-feedback environment in which the child's normal routine is prohibited and a new, permitted one is heavily practiced.

We are currently testing a form of pure prohibition instruction. Consideration of Brown and VanLehn's theory of the origin of bugs leads us to expect that pure prohibition instruction may eliminate bugs in the short run, but the bugs may reappear later or new buggy algorithms may replace them. Further, we expect that little transfer or invention of the kind shown by our three initial subjects will occur. If, as Brown and VanLehn argue, bugs derive from children's active attempts to invent procedures for dealing with new situations, then prohibition of wrong moves cannot by itself provide any basis for successful rather than maladaptive inventions. An instructor can prohibit an operation in writing, but if no new information is offered to the system, then the system will have no basis for responding with an appropriately modified set of operations.

Enrichment

For these reasons, we are inclined to believe that something more than prohibition took place in our mapping instruction, that there was some kind of enrichment of the knowledge structure that permitted the children to build some new connections and thus make sense of situations newly encountered some weeks

---

[1]A few bugs are not strictly prohibited by the blocks. Of the bugs shown in Fig. 10.2, Zero-Instead-Of-Borrow is physically possible. If prohibition were the only way in which mapping affected writing performance, one would, therefore, expect this bug to remain intact even after extensive mapping practice. This has not yet been tested in our work, as none of the children taught thus far had these particular bugs at the outset.

later. If we imagine that the children began the mapping instruction with an already well-developed justification for the operations in block subtraction based on the semantics of place value and subtraction, our theory will need to show how this semantic knowledge can be incorporated into the knowledge structure for written arithmetic as a result of mapping activity. Figure 10.4 shows the mapping links between a child's semantic knowledge and a written procedure.

The left side of Fig. 10.4 schematizes the semantic knowledge relevant to subtraction that is instantiated in the blocks routine. There are two aspects to this knowledge: (1) a goal structure for blocks subtraction and (2) knowledge about the kinds of exchanges allowed within the base-ten system. Only knowledge directly relevant to borrowing is shown, not the underlying knowledge about how base-ten information is coded in blocks. The goal structure that controls the exchange/borrowing process is shown at the far left. Constraints on each goal are shown to its right by a single arrow pointing to the goal. In each case a subgoal is generated in order to meet the goal. The main goal (*A*) is to remove the bottom quantity from the top—which, it is assumed, has been represented in blocks according to the rules specified in the base-ten code. This main goal requires that there be at least as many objects in the top of the active column as in the bottom. Otherwise it is necessary to get more blocks for the column (*B*), and this subgoal must be achieved without changing the quantity as a whole (i.e., while maintaining equivalence). This can be accomplished through trading. The dotted single arrow between getting more (*B*) and trading (*C*) reflects the finding, described earlier, that some children need prompting before they initiate trades to solve concrete subtraction problems. Trades are constrained by the requirement that there be both an increment and a decrement, and by the requirement that the trade be "fair" (ten for one). The ten-for-one relationship built into trades means that the only way to decrement from the hundreds column and increment in the units column is to do a double trade (*D*).

The right side of Fig. 10.4 represents the components of the written algorithm stripped down to include only the steps involved in borrowing. The paths for problems containing zeros and not containing zeros are shown in the left branch and the right branch respectively. The operations in the two branches can be done in several orders, but those shown are the orders frequently observed in algorithmic performance. Omission of one or more operations will produce buggy algorithms.

Operations-level mapping between blocks and written procedures would link each operation in the written procedure to a corresponding operation in blocks. The double arrows in Fig. 10.4 represent the kinds of links that mapping might be expected to produce. First, the top-level goal of the blocks routine (*A*) becomes linked to the written algorithm. A particular effect of this is to motivate testing (at *a*) whether the top number in a column is greater than or equal to the bottom, thus reducing the probability that this crucial step will be omitted. The next goal (*B*), getting more blocks, may also be linked to the written algorithm as indicated by the dotted box below *a*. Mapping of the trading goal (*C*), with its

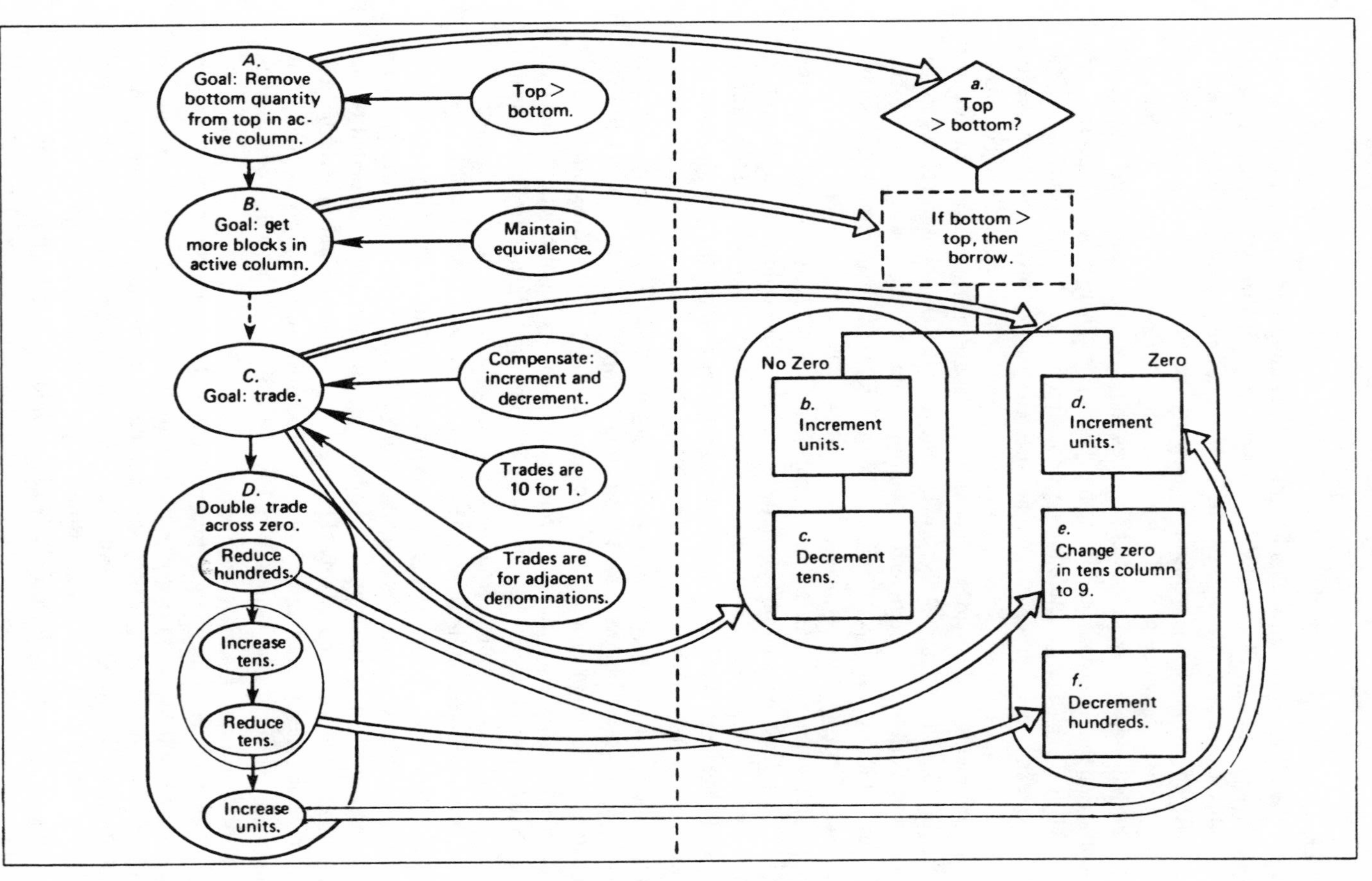

FIG. 10.4. Conceptual links facilitated by mapping.

constraints, can be expected to produce a "chunking" of incrementing and decrementing in one's mental representation of the written algorithm. This chunking in the algorithm is shown by the circles that now enclose steps *b* and *c* and steps *d, e,* and *f*. The chunking ensures that incrementing and decrementing are seen as complementary actions. For problems with zeros in the top number, the operations on zero are tied to the incrementing and decrementing operations that surround them. In addition to organizing the steps of borrowing into chunks—thus probably making omissions of individual steps less likely—the mapping has the effect of imposing the trading constraints (compensation, ten-for-one, and adjacency) on the written algorithm. Borrowing becomes, in effect, understood as an analog of trading, and the rules of trading become available as justification or explanation of the steps in borrowing. The links from the subop-erations within *D* to specific steps in the *d–e–f* sequence suggest how this may work. Probably the most important aspect of the detailed mapping is that it "explains" why 0 should be changed to 9—something that mystifies many children, according to our interview data.

This kind of enrichment of the knowledge structure can be expected to make the various steps in a correct writing algorithm easy to remember, because they are embedded in a knowledge network that justifies and explains them. Perhaps more important, the enriched knowledge structure should provide a basis for modifying or building routines that would make future repairs on subtraction algorithms likely to produce correct rather than buggy routines. The effect of enriched knowledge might be expected both at the point of *generating* possible operations and at the point of *testing* them. The semantic knowledge available from mapping would suggest potential incrementing and decrementing moves. Constraints that are part of the semantic knowledge would also serve to block possible incorrect operations that children might generate. The parallel between this informal account of the effects of mapping and Brown and VanLehn's (this volume) formal account of the origin of bugs suggests the possibility of a formalization of enrichment as an extension of their present repair theory. What is needed to guide such an effort is direct observation of the invention and repair of algorithms by children.

REFERENCES

Brown, J. S., & Burton, R. R. Diagnostic models for procedural bugs in basic mathematical skills. *Cognitive Science,* 1978, *2,* 155–192.

Young, R. M., & O'Shea, T. Errors in children's subtraction. *Cognitive Science,* 1981, *5*(2), 153–177.

11 General Developmental Influences on the Acquisition of Elementary Concepts and Algorithms in Arithmetic

Robbie Case
Ontario Institute for Studies in Education
University of Toronto

In this chapter, I advance four basic hypotheses:

1. The development of cognitive abilities is parallel across various domains of activity.
2. A major reason for this parallel is that cognitive development is constrained by the growth of central processing capacity.
3. Mathematics abilities exhibit the same pattern of development as other abilities.
4. This pattern of development has important implications for the design of effective instruction.

CROSS-DOMAIN PARALLELS IN COGNITIVE DEVELOPMENT

When the types of operations that children develop at different ages are analyzed, certain interesting parallels emerge across a wide variety of domains.

Development During the Preschool Period

Substage 1: Operational Consolidation (1–1½ Years). Between the age of 1 and 1½ years, children assemble a variety of new operations. Two of these are illustrated in Table 11.1. In the domain of language, children use their first words to request assistance from an adult. In the domain of constructive play, children build their first toy objects out of blocks. Normally the first block construction to appear is the tower, which children either discover on their own or see demonstrated by an older child.

At first glance, these two operations might appear to have relatively little in common. One is social in origin; the other is not. One involves physical activity; the other does not. Although the two operations do have their origins in quite different domains, however, they share two important features as well. They both require the child to form an internal representation of a relation between two external objects, and they both require the child to execute a motor response that will produce this relationship.

Consider the child's first verbal requests. These requests are normally for an object that is just out of his reach. Moreover, they are addressed to an adult who is very close at hand. As Bates (1976) has pointed out, this first use of language requires that the child see the adult as a means to an end. In effect, the child must encode a desired relationship between the adult and the object that is out of reach (*A* can bring *B*) and execute a complex motor response (speech) that will produce this relationship. In the domain of constructive play, the situation is quite similar. In order to build a tower, a child must encode a desired relationship between two blocks (*A* can rest on top of *B*) and must then execute a complex motor response that will produce this relationship.

If these two tasks were the only ones children mastered at this age level, the parallel in their structure might be dismissed as a coincidence. However, we have recently studied 10 different operations in four different domains. Moreover, we have found that all the operations that children master during this time period share this same structure (Case & Khanna, 1981). This structure might be characterized as follows: Sensorimotor response $A \rightarrow$ (relation) B. Accordingly, we have begun to refer to the preschool period as the period of *relational operations*.

TABLE 11.1
Cross-Domain Parallels in Children's Development

		Achievement	
Substage	Age	Language	Constructive Play
1.	1–1½ yrs	Johnnie Ball	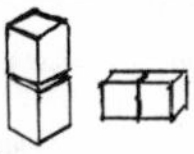
2.	1½–2 yrs	Want ball Johnnie ball My ball	
3.	2–3 yrs	Johnnie want my ball	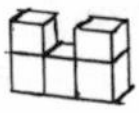
4.	3–4½ yrs	Johnnie want the big ball in the kitchen.	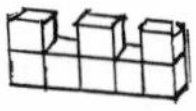

Substage 2: Operational Coordination (1½–2 Years). Between the age of 1½ and 2, few genuinely novel operations appear. Rather, children become capable of coordinating two operations of the sort that they mastered earlier. In the area of language, they begin to generate two-word utterances rather than single words. In the domain of block building, they begin to build objects like the train in Table 11.1. Note that the train involves two relations of the type mastered at the first substage. One relation is "on top of." The other relation is "beside." A similar progression is observed for most of the other domains we have studied.

Substage 3: Operational Coordination (2–3 Years). Between the ages of 2 and 3, the complexity of children's operational structures increases once again. In the domain of language, children now generate utterances made by pairing the simpler phrases they mastered earlier (for example, [baby want] [more juice]). As a consequence, a good proportion of their utterances are now complete sentences. In the domain of constructive play, children now master constructions like the arch or the fort. Once again, both these constructions involve coordinating two elements that were mastered individually at an earlier stage. The elements in question are illustrated in Fig. 11.1.

Substage 4: Bifocal Coordination with Elaboration (3–4½ Years). During the fourth substage, children's intellectual structures are elaborated by the addition of another element. In the domain of language, the result is that children now generate sentences containing subordinate phrases or clauses attached to one of the elements in the main clause. In the domain of block building, children now construct elaborate structures like those in Fig. 11.1. Note that these structures have the same basic form as those at the earlier substage with the addition or repetition of an element. Once again, a similar elaboration is observed for most of the other eight tasks we have studied.

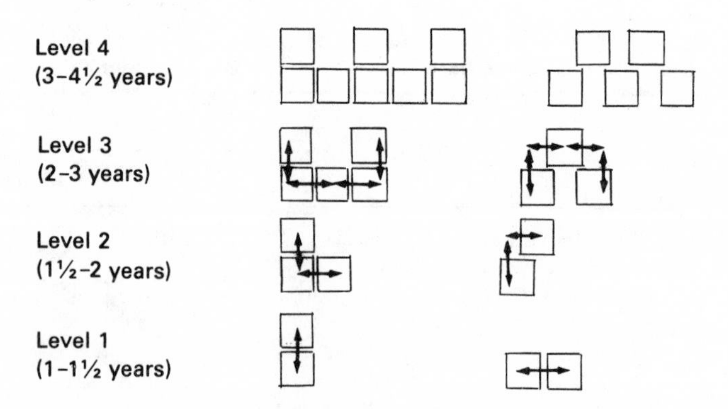

FIG. 11.1. Structures mastered by preschoolers in block building and their component relations.

Development During the Elementary School Period

During the elementary school years, we observe a progression very similar to that seen in the preschool period. The primary difference is that the operations being assembled and coordinated are of a higher level.

Substage 1: Operational Consolidation. During the same substage that culminates children's relational development, they begin to consolidate a new type of operation that serves as the foundation for their development during the elementary school period. In this operation, children evaluate their perceptual world along some polar dimension (e.g., big, small). They then use this evaluation for making predictions about the occurrence or nonoccurrence of some significant event (e.g., sinking, floating). Two examples we have looked at in some detail involve a balance beam and a story about a birthday party. In the balance beam example, children at ths substage globally evaluate the objects that are placed on a beam and decide which one is heavy and which one is light. They then predict that the side with the heavy weight will go down (Liu, 1981). In the birthday party example, children evaluate the presents received by two children and decide which present is nice and which one isn't. They then predict that the child who received the nice one will be happy, and that the child who received one that wasn't nice will not (Marini, 1981). On a wide variety of other tasks—both those we have developed and those developed in other laboratories—a similar sort of operation is seen to emerge during the same time period (Case, in preparation). The structure of these operations might be characterized as follows:

Dimension A (pos., neg.) → Event B (pos., neg.)

Accordingly, we have begun to refer to this period as the period of *dimensional operations*. We have also analyzed the internal structure of dimensional operations and have shown that they are assembled from four component relational operations (Case, in preparation).

Substage 2: Operational Coordination (4½–6½ Years). During the second substage, children coordinate their evaluation of one dimension with that of another. For example, on the balance beam problem, they now coordinate their evaluation of weight with an evaluation of number, thus employing the first of the quantitative "rules" studied by Siegler (1976). Using this rule, children predict that the side with the greater number of weights will go down, in situations where appearance gives little or no cue as to relative weight. Similarly, in the birthday party example, children now coordinate their evaluation of the quality of each child's gift with an evaluation of quantity. Consequently, they predict that the child who gets the greater number of presents will be happier if the general type of present is the same. Once again, a similar sort of operational coordination is observed across a wide variety of domains. Siegler (in press) has presented data in four different domains. We ourselves have reviewed data from another six (Case, in preparation).

Substage 3: Bifocal Coordination (7–8 Years). During the third substage, children's focus expands to include an additional operation or operational pair. In the balance beam task, children now focus on the number of distance units of each stack of weights from the fulcrum, as well as the number of weights in each stack (Siegler, 1976). In the birthday party task, they now begin to focus on the number of gifts (e.g., marbles) each child wanted, as well as the number actually received. If two children received the same number of marbles, but one wanted more, they now predict that the child who wanted more will be less happy (Marini, 1981). Similar trends are once again presented on a wide variety of other tasks from concept learning through analogical reasoning (Case, in preparation).

THE ROLE OF CENTRAL PROCESSING CAPACITY

The analyses in the previous section are quite global. Nevertheless, they suggest an interesting hypothesis: The child's transition from one stage and substage to the next in any given domain depends not just on experience in that domain, but on the growth of some central coordinating or processing capacity. This hypothesis is by no means a new one. It was first suggested at the turn of the century by J. M. Baldwin and was mentioned in passing several times by Piaget. It has also formed the cornerstone of the theory of development proposed by Pascual-Leone (1970).

My own theory of development is an extension of Pascual-Leone's. I have suggested that a child can maintain one operation in his central processing space at the beginning of any major period, and that this number increases arithmetically with each successive substage. I have also suggested that the ability to coordinate a certain number of elements at one stage is a prerequisite for assembling the operations of the next higher stage.

The reason I have suggested an arithmetic increase in capacity perhaps deserves comment. The maximum number of external elements that a child can coordinate often increases geometrically. However, because of "chunking," not all this increase in external performance need be attributed to an increase in internal capacity. For example, the child who can repeat "baby wants" may indeed need a capacity of 2. However, this child should be able to repeat [baby wants] [more juice] with a capacity of only 3, inasmuch as the second two words can be stored as one chunk while the first two are processed. A more detailed explication of the assembly and "unpacking" of such chunks is presented elsewhere (Case, in preparation).

Over the past few years, my students and I have conducted a number of studies to investigate the role that central processing capacity plays in children's cognitive development. Our findings to date may be summarized as follows:

1. The tests of central processing capacity we have devised have shown the same norms we predicted in our theoretical analyses. For example, when preschoolers are asked to repeat sets of unrelated words, they can repeat one such word by the age of 1½, two such words by the age of 2, three words by the age of 3, and four by the age of 4½. Similarly, if they are asked to imitate a relational motor operation, they can imitate one such operation at 1½, two at 2, three at 3, and four at 4½ (Liu, 1981). The same trend appears at a higher level during the dimensional period. If children are asked to count an array of objects and to remember the total, they can do this for one array by the age of 4½, two arrays by the age of 6, and three arrays by the age of 8. Similarly, if they are asked to perform a nonverbal dimensional operation such as spatial localization, they can localize one object in a grid at age 4½, two at age 6, three at age 8, and four at 10–11 (Case & Kurland, 1980). During the two major periods of development under consideration, then, the growth of processing capacity as measured by our tests corresponds with the theoretical values derived from our prior theoretical analyses.

2. A second type of data we have gathered is correlational. The correlations between measures of quantitative capacity such as those mentioned in the foregoing paragraph and children's domain-specific cognitive development are quite substantial. Values as high as .85 have been found for the relational period (Liu, 1981) and values as high as .65 for the dimensional period (Case & Watson, 1979).

3. A third type of data involves training. If one takes children who have not yet learned the bifocal strategy on the balance beam and divides them into groups with a high and low quantitative capacity, one discovers that the subjects with high capacity profit from instruction very rapidly, whereas those with a low capacity profit very slowly, if at all (Case & Watson, 1979). Similar results have been shown for a variety of other developmental tasks (Case, in preparation).

4. A fourth type of evidence involves the experimental reduction of adults' functional processing capacity. The method by which this is done is too detailed to elaborate here. What one finds under these circumstances, however, is that adults form a structure of the same complexity as children when their processing capacity is of the same magnitude, providing that the task with which they are confronted is genuinely novel.

Together, these four lines of evidence suggest that the development of central processing capacity plays an important role in determining the rate of children's cognitive development in a variety of domains. That is not to say that it is the only factor that plays this role, of course. It is merely to say that it is one such factor, and that it works by limiting the number of internal elements children can attend to simultaneously.

If central processing capacity exerts this kind of regulatory role on the rate of development, a reasonable question to ask next would be: What determines the rate of growth of processing capacity itself? Although space does not permit an adequate answer to this question in this chapter, two conclusions we have reached in our recent studies are worth mentioning. First, the basic factor that controls the growth of processing capacity is operational efficiency. In effect, a child's total processing capacity may be thought of as fixed, and the amount available for storage may be seen as a function of the proportion for this total capacity that must be devoted to executing basic operations (Case, Kurland, & Goldberg, in press). A second conclusion is that operational efficiency is a function of both maturation and practice (Kurland, 1981). Stated differently, practice can improve a child's operational efficiency in a given domain, but only up to the limit imposed by maturation.

PARALLEL TRENDS IN THE AREA OF MATHEMATICS

If the above view of development is correct, it follows that children should evidence the same stagelike progression in the area of mathematics as they do in other domains.

Development During the Preschool Period

A development of particular importance for mathematics takes place during the preschool period. This is the development of the ability to count. We have recently examined children's counting and found that its development can be traced back to primitive relational operations that first emerge at the age of 1–1½.

Substage 1: Operational Consolidation. One of the basic components of children's first attempts to count is the transfer of an object from one pile or location to another. In the context of the analyses presented earlier, this is a relational operation. It involves encoding a desired relationship between two external objects (e.g., marble inside box) and then generating this relationship via a sensorimotor operation.

Another basic component of the child's first attempts at enumeration is producing a symbolic label while transferring an object from one location to another. As mentioned in the first section, children's first words emerge during this same stage. With a little practice, they can mimic most simple words uttered by an adult, including the number names. Although they can say the word "one" in a word game, however, and although they can transfer a marble from a pile into a box, they cannot yet coordinate these two operations with each other.

Substage 2: Operational Coordination (1½–2 Years). By the age of 1½–2, children become capable of precisely this sort of coordination. If the experimenter says a number name while putting a marble in a box, they will now mimic both components of this activity. Alternatively, if an experimenter says two number names in a row, or pats two objects in a row, they will mimic this. As yet, however, they still cannot coordinate these operational pairs with each other. Hence, they cannot assign a sequence of number names—even as simple as 1, 2—to a sequence of transference or pointing acts.

Substage 3: Bifocal Operational Coordination (2–3 Years). The bifocal coordination necessary for combining these pairs emerges at about the age of 2. At about this age, children begin to be able to imitate an experimenter in counting two objects, and by the end of the substage they have extended this competence to sequences as large as five or more objects (Gelman, 1978). This is the first substage, then, at which children's attempts at counting might be said to be completely successful, as they can now map a conventional sequence of symbols onto a sequence of transferring or pointing acts. There are still certain counting tasks, however, with which they have problems, and which they do not master until the final substage of the period.

Substage 4: Elaborated Bifocal Coordination (3–4½ Years). During the final substage, children become capable of counting a particular type of object in a line of objects (e.g., counting just the girl dolls in a line of boys and girls). They also become capable of counting a set of objects that are not arrayed in a line, and where careful track must therefore be kept of which objects have been counted and which have not. Note that neither of these abilities involves a change in the basic structure of the counting algorithm. However, both involve an elaboration: the former an elaboration for classifying objects as part of the set to be counted, the latter an elaboration for classifying objects as already having been counted.

Development During the Elementary School Years

During the elementary school years, children appear to go through a sequence of developments in the area of addition that parallels the development they went through previously in the area of counting, but at a higher level.

Operational Consolidation (3–4½ Years). The same period that sees children applying their counting routine to difficult enumeration problems sees them consolidating their first procedure for adding two sets of objects. During this age range, if children are asked to find the sum of two numbers (e.g., 2 + 3), they proceed in a fashion that has been called "counting all" (Fuson, this volume). They count out two tokens. Then they count out three tokens. Then they put all the tokens together and recount them. Note that the basic operation this addition routine requires is counting, and that counting may therefore be said to play a pivotal role in the development of mathematical abilities.[1] From the viewpoint of the relational stage, counting involves the coordination of several different relations and represents the culmination of several years of cognitive growth. From the viewpoint of the dimensional stage, counting is a fundamental operation, responsible for abstracting one of the most crucial dimensions on which further development depends, namely the dimension of number.

A second point to note is that children who add in the manner described as "counting all" never have to coordinate one counting operation with another. They never have to store the product of one counting operation while executing a second. Nor do they ever have to anticipate the product of a second operation while executing a first. In effect, then, these children's procedure for addition enables them to perform the task within the constraints imposed by the limited central processing capacity they have at this point in their development.

Operational Coordination (4½–6½ Years). During the second substage, children's procedure for addition changes. When asked to add two numbers, children now count out the first set, count out the second set, and then "count on," recounting the first set from where they left off counting the second (Fuson, this volume). Note that counting on implies an understanding that to recount both sets would be uneconomical, as one would get exactly the same result in recounting the second set as one did the first time. This might require a dimensional coordinating capacity of 2, because—in order to acquire it—a child would have to be able to store the product of one counting operation while executing a second. Once again, then, we see a correspondence between the central processing capacity children are known to have at a certain substage and their level of mathematical development.

Bifocal Operational Coordination (6½–8 Years). During the third substage, children develop methods of addition that enable them to perform the task without recounting either set. In effect, they count all the first set and then count on from there, while simultaneously making their first count of the second set (e.g., "one, two, three (pause), four–one, five–two." The exact age at which children make this shift to simultaneous counting and the exact procedures they use are not yet clear (see Fuson, this volume, for relevant data). However, regardless of the precise method used, it seems clear that the new procedures place an additional load on the child's central processing capacity.

---

[1] It plays a similar role in the development of many other abilities as well (cf. Case, in preparation).

The analyses I presented in this section were quite global. Nevertheless, they should be sufficient to make my general point: There is a parallel between the sort of development observed in the area of mathematics and that observed in other domains. This parallel may be seen: (1) in the level of operation that children employ; (2) in the stagelike manner in which these operations are coordinated into ever more sophisticated and economical procedures; and (3) in the ability of the child's processing capacity to handle the quantitative load that these increasingly complex procedures entail.[2] I turn now to a consideration of the instructional implications of my view of cognitive development in general and of mathematical development in particular.

IMPLICATIONS FOR INSTRUCTION

There would appear to be at least three implications for instruction of the view of development I have presented.

1. Match Instruction to Students' Current Developmental Level

If students pass through a natural sequence of substages en route to full mastery of any mathematical concept or skill, it seems reasonable to suggest that instructors should: (1) determine what these substages are; (2) determine what substage their own students are actually at; and (3) design their instruction so that it will bring the students from their current level to the desired level as efficiently as possible.

These three suggestions are of course not new. They have been made by many other educators, particularly those who have been interested in Piaget's theory (Hunt, 1961). Three dilemmas have plagued Piagetian attempts to translate these suggestions into practice, however. These are: (1) the difficulty of determining what structure might be relevant for the acquisition of a given mathematical concept; (2) the difficulty of assessing whether or not students have acquired this structure; and (3) the difficulty of suggesting what sorts of activities would be most likely to bring students from one of these levels to the next (Case, 1978b).

The aforementioned problems are less serious within the context of my theory, as structures are seen as being domain specific, linked only by common working memory demands:

1. To determine the structures or strategies that define levels in any given domain, one need only examine the performance of several different age or experience levels within that domain. A procedure for doing so has been spelled

---

[2]More detailed analyses are available in Case, 1979b.

out elsewhere (Case, 1978a). Basically the procedure is quite simple. One examines children's responses to the target task and interviews them to determine how they arrived at these answers. Then one invents new tasks to test hypotheses about the children's underlying strategy. Examples of this approach are available in Siegler's (1976) work on the balance beam or Fuson's (this volume) work on counting strategies.

2. To determine the level of one's own students, one simply presents the problems that were developed in the first place and uses the children's responses to determine which of the various strategies or levels of understanding they are currently employing.

3. Finally, to facilitate children's progress from one level to the next, one simply presents them with problems one level beyond those they can currently handle, in a context where they can determine for themselves that their current strategy yields an incorrect answer. One then helps them to see what feature of the new problem they have ignored, and to incorporate a new component into their current strategy for dealing with it. A detailed example of this sort of approach in the domain of mathematics has been presented by Gold (1980; Case, 1980).

The preceding three steps for matching instruction to students' current developmental level can be of particular importance for mathematical skills that are known to be difficult to teach, and where the majority of students are known to give the same wrong answer when presented with sample problems. Such a case occurs in the area of early addition, when problems such as 5 + ______ = 8 are introduced. Children normally give one of three answers to this problem. The first is to write in one of the numerals in the problem (i.e., 5 or 8). The second is to write in their sum (i.e., 13). The third is to write in their difference (i.e., 3). Obviously, students who write in the difference have attained the correct answer and do not need further instruction. Obviously, too, children who write in one of the given numerals have not appreciated that the symbols indicate the operation to be one of addition. What about students who write in the sum? These students apparently have not appreciated the significance of the order in which the digits are written. Stated differently, they have not understood the basic nature of an equation, and so they treat all problems with two numbers and a plus sign as simple addition tasks. If one examines the strategies that the children use to give this answer, one finds that they are identical to those used by the children who give the correct answer, in the sense that they also count on. However, they are different in that those who succeed count on from 5 to 8, noting the number of tokens or fingers that they have used in doing so, whereas those who fail count on from 8, until they have gone 5 past it. Understanding the strategy that is used by both groups—those who pass and those who fail—permits one to bridge the gap quite easily, both with very young children (i.e., kindergarten) and with math-disabled children (Case, 1979a; Gold, 1979).

2. Minimize the Processing Load of Developmental Progression

Developmentalists generally agree that the strategy of presenting children with a problem one level beyond their current level of capability is an optimal procedure for promoting transition to a higher level. They are not always optimistic, however, about the possibility of success, particularly if they see the limiting factor in development as some "underlying structure of the whole." In my own theory, the limiting factor is the size of the child's processing capacity. Thus, one can be quite optimistic about the possibility of facilitating transition to a higher level, if steps are taken for circumventing such capacity limitations. The second general implication of my view of development, then, is that one should minimize the processing load during the developmental transition.

Several different procedures for minimizing this load may be suggested:

1. Teach a strategy that has a light load associated with it to begin with. Teaching the missing-addend problem via a counting-on strategy rather than a "simultaneous counting" strategy or a "reverse the sign" strategy would be an example of this sort of approach.

2. Minimize the number of new elements that are added in any one curriculum unit. As an illustration of this approach, suppose that a child already had a successful strategy for adding real objects (e.g., the count-on strategy), but was having difficulty learning to deal with symbolically stated problems. The transition to this level of functioning might be done as follows. First, one could introduce problems where the only symbols were the plus and equal signs. The child might be asked to write in the appropriate answer on sheets of problems such as this:

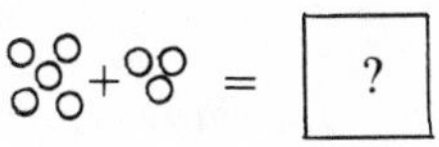

ooooo + ooo = [?]

One could then move on to problems where a numerical symbol was used as well. Perhaps one could say that someone had put some money in a piggy bank, and had written down how much was there. The problem is to figure out how much the person will have if some more pennies are put in. One could then ask the child to write in the appropriate answers on a sheet of problems like the following:

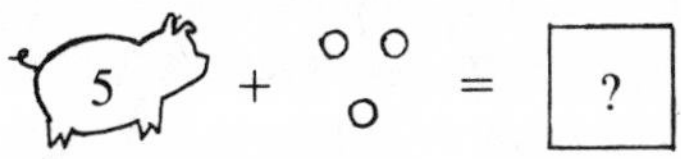

(5) + ooo = [?]

Next one could move on to the problems of this sort:

5 + 3 = ?

and finally to problems having the standard format.

3. A third way to minimize capacity load is to provide sufficient practice at each problem level that the child's choice of procedure for each type of problem becomes automatic. The reason for this is plain. If a child still struggling to master one level of task is pressed to a higher level, the child will have less free-processing capacity for dealing with the new one.

4. A fourth procedure is to utilize mnemonic aids. Children asked to solve the missing-addend problem in their head, for example, will find this more difficult than if they are allowed to use their fingers, as fingers serve as an extra storage slot for keeping a record of the units counted in moving from one number to the next.

5. A fifth procedure is to devise all visual props so that the relevant cues are highly salient. Extracting a relevant cue from an irrelevant context is another activity that can take up free-processing capacity and decrease a child's maximum level of functioning.

6. A sixth procedure is to eliminate any unnecessary conceptual or computational complexities when a task is first being introduced. In introducing double-digit addition, for example, one should first use small numbers with which the child is already efficient and only later graduate to larger ones.

7. A final procedure is to eliminate any verbal or other complexity associated with the initial presentation of the task. Unnecessary complexity should be removed from the teacher's verbal presentation and the task itself.

3. Insure the Child's Basic Operations Are as Efficient as Possible

At the end of the second section, I mentioned that children's central processing capacity—or at least that proportion of it that is functionally available—grows with age, due to increasing operational efficiency. Although there is a maturational component to this increase, considerable practice is necessary in order to insure that the ceiling level set by maturation is actually realized. A third implication of my theory, then, is that sufficient practice in basic operations should be presented to insure that children realize their potential.

For operations such as counting, children probably encounter sufficient practice in their natural environment to insure an optimal growth of efficiency. We have found that very little additional improvement in efficiency takes place in first grade, even with the addition of 20 minutes per day of counting practice

(Kurland, 1981). For operations such as addition and subtraction (or later multiplication and division), it is by no means certain that practice is unnecessary. Some children may perform such operations far less efficiently than their peers, even though they use the same basic strategy. Similarly, whole classes of children may perform these operations less efficiently than they could with sufficient practice. The simplest way to check children's operational efficiency is to get them to perform a series of basic operations as fast as possible and then divide their total time by the number of items completed. This value can then be compared to national or locally collected norms, and children who fall considerably below the norm can practice until they reach it, or at least until they reach an asymptotic value. We have found that time charts provide good motivation for this activity, as children can see themselves progressing day by day (Kurland, 1981).

Although the ultimate concern in mathematics instruction should always be to produce conceptual understanding as well as efficiency in basic operations, the interaction between the two should not be forgotten. Inadequately automatized operations at one level will take up more processing capacity than is necessary and thus make the mastery of the next higher level concepts or algorithms more difficult.

REFERENCES

Bates, E. *Language and context: The acquisition of pragmatics*. New York: Academic Press, 1976.

Case, R. A developmentally based theory and technology of instruction. *Review of Educational Research,* 1978, *48,* 439–469. (a)

Case, R. Piaget and beyond: Toward a developmentally based theory and technology of instruction. In R. Glasser (Ed.), *Advances in instructional psychology*. Hillsdale, N.J.: Lawrence Erlbaum Associates, 1978. (b)

Case, R. Implications of developmental psychology for the design of instruction. In R. Glaser, A. Lesgold, J. Pellegrino, & J. Fokkema (Eds.), *Cognitive psychology and instruction*. New York: Plenum, 1979. (a)

Case, R. *Learning, maturation and the development of computational strategies in elementary arithmetic*. Paper prepared for the Wisconsin Conference on the Initial Learning of Addition and Subtraction Skills. Racine, Wisconsin, 1979. (b)

Case, R. *A developmentally based approach to the problem of instructional design*. Paper presented at the NIE Conference on Teaching and Thinking Skills. Pittsburgh, October 1980.

Case, R. *Intellectual development: A systematic reinterpretation*. New York: Academic Press, in preparation.

Case, R., & Khanna, F. The missing links: Stages in children's progression from sensorimotor to logical thought. In K. W. Fischer (Ed.), *New directions for child development*. San Francisco: Jossey–Bass, 1981.

Case, R., & Kurland, D. M. *Construction and validation of a new test of children's M-space*. Unpublished manuscript, University of Toronto (OISE), 1980).

Case, R., Kurland, D. M., & Goldberg, J. Operational efficiency and the growth of short term memory span. *Journal of Experimental Child Psychology,* in press.

Case, R., & Watson, R. *Teaching young children to take account of two variables on the balance beam task*. Unpublished manuscript, 1979.

Gelman, R. Counting in the preschooler: An analysis of what does and does not develop. In R. Siegler (Ed.), *Young children's thinking: What develops?* Hillsdale, N.J.: Lawrence Erlbaum Associates, 1978.

Gold, A. P. *Cumulative learning versus cognitive development: A comparison of two different theoretical bases for planning remedial instruction in arithmetic*. Unpublished doctoral dissertation, University of California–Berkeley, 1979.

Gold, A. P. A developmentally based approach to the teaching of proportionality. In *Proceedings of the Fourth International Conference for the Psychology of Mathematics Education,* Lawrence Hall of Science, Berkeley, 1980.

Hunt, J. McV. *Intelligence and experiences.* New York: Ronald Press, 1961.

Kurland, D. M. *The effect of massive practice on children's operational efficiency and short term memory span*. Unpublished doctoral dissertation, University of Toronto (OISE), 1981.

Liu, P. *An investigation of the relationship between qualitative and quantitative advances in the early cognitive development of preschool children*. Unpublished doctoral dissertation, University of Toronto (OISE), 1981.

Marini, Z. *The relationship between children's understanding of affect and their cognitive development*. Unpublished master's thesis, University of Toronto (OISE), 1981.

Pascual-Leone, J. A mathematical model for the transition rule in Piaget's developmental stages. *Acta Psychologica,* 1970, *32,* 301–345.

Siegler, R. S. Three aspects of cognitive development. *Cognitive Psychology,* 1976, *8,* 481–520.

Siegler, R. S. *Developmental sequences within and between concepts*. Society for Research in Child Development. Monograph, in press.

12 The Structure of Learned Outcomes: A Refocusing for Mathematics Learning

Kevin Collis
University of Tasmania

The purpose of this chapter is to describe a model for categorizing students into developmental levels on the basis of their responses to particular content areas and to illustrate the model's use with mathematics. The model grows out of a shift in the focus of research related to how children respond to particular tasks. In producing a response, two related phenomena appear to be involved. First, there is an underlying phenomenon that defines the individual's cognitive limits termed the *hypothetical cognitive structure* (e.g., Piaget's stages of cognitive development). The second phenomenon, which is a function of the hypothetical cognitive structure and of experience in a content area, is termed the *structure of the learned outcome* (Collis & Biggs, 1979). The former is a fixed structure at any particular time and is presumably unalterable by what teachers do (at least in the short term), whereas the latter is flexible and alterable by instruction. In fact, the latter could be used by the teacher to guide the way in which lessons and programs are designed. What is argued in this chapter is that the focus of research should shift away from a response simply implying a stage of cognitive development to consideration of the *quality* of each individual response per se.

BACKGROUND

To appreciate this refocusing and its importance let me briefly trace the evolution of my ideas on cognitive development and how those ideas led to the development of this response model.

First of all, the root of my ideas is firmly planted in my own early experiences teaching mathematics in the classroom. A series of quite practical teaching and curriculum problems needed to be solved at that time. Attempts to solve these problems led to what might be described as a *descriptive* phase in my thinking where the problems of mathematics learning were related to a Piagetian developmental framework (Collis, 1969, 1971, 1973, 1974, 1975a, 1975b). This was followed by an *explanatory* phase in which the developmental ideas were related to certain information-processing concepts (Collis, 1980a, 1980b). In my most recent work (Collis & Biggs, 1979), I have moved away from further detailed analysis of the developmental stages with their implications for Hypothetical Cognitive Structures (HCS) to a consideration of the Structure of the Learned Outcome (SOLO). The latter is a response model that relates the ideas developed in earlier work to the practical problems faced by teachers and curriculum designers.

In 1975 Professor J. B. Biggs of the University of Newcastle and I set out to find examples of levels of cognitive development in the major content areas in high school and late primary school so that these examples could be made available to teachers. Our aim was to describe qualitatively different levels of judgment and reasoning ability as they apply to the teaching of different subjects so that a teacher could gauge each student's learning and thinking.

However, as we began the literature review several problems soon became apparent, three of which were:

1. The criteria used by various researchers for determining different developmental stages differed considerably across content areas. Inverse operations, for instance, although quite clearly useful in logico-mathematical tasks, did not seem as significant in other areas such as English or history.
2. When testing subjects on items from different content areas the well-known Piagetian concept of decalage was found to be very much the rule rather than the exception. This finding was at odds with both the developmental approach and the idea of developmental stages. The same student was sometimes observed to vary three and four stages in the same content area on different occasions. Age ranges for typical responses seemed too gross a measure to accommodate the traditional stage development scheme.
3. Another puzzling feature that arose was inconsistencies in the same student's response upon retesting on the *same item*. A student might respond at what would be termed a middle concrete level and then, when retested some time later, give responses one or two levels higher, or even lower.

Such considerations forced us to question the notion of categorizing students into developmental levels on the basis of their responses to particular items in particular content areas. To achieve our purposes we needed to shift away from a model in which a response implies a stage of development and focus instead on the quality of the response at a particular time. Like a score on a classroom test this categorization might vary from day to day, depending on factors other than stage of cognitive development—factors such as emotional state, physical state, and motivation.

The taxonomy we have devised appears to offer a useful way of analyzing what is meant by quality of response.

In summary, a student's response can be categorized in two domains. The growth of thought proceeds from sensorimotor actions through various stages to the complex structure of formal operations. These internalized stages are called the *hypothetical cognitive structure* (HCS) (Biggs, 1980), which is a property of the individual. On the other hand, educators know that students' responses to a learning task proceed from simple repetitions of the given to complex structures moving beyond the particular information offered—structures that may display elegance, deftness, surprise, and beauty. This domain is called the *structure of observed learning outcomes* (SOLO), which is task specific.

These points may best be illustrated by enumerating the five stages[1] of cognitive development and relating these to the levels in the SOLO Taxonomy.

Preoperational Stage. This is the stage at which, in the Piagetian model, responses are nonlogical or prelogical. They are based on a lack of comprehension: tautology, an irrelevant association, or denial of the data. The equivalent level in the SOLO Taxonomy is called *prestructural.* Here the responses indicate that the child has no real feeling for what one would call mathematics.

The Early Concrete Operational Stage. An elementary basis for a concrete logic of classes, differences, and equivalences now exists. Conservation, which involves elementary use of the reversibility principle, seems well established and there is a basis upon which to develop some sort of logical structure including a mathematical structure. The equivalent level in the SOLO Taxonomy is termed *unistructural.* At this level children demonstrate a need to close quickly any operation given to them and find it difficult to give meaning to expressions that have more than one operation with small numbers. In solving any problem they tend to go forward from the starting point and then only one step. For example, in solving the equation $y + 4 = 7$ they will give the response 3; their reason involves some sort of "counting-on" procedure. Typically they will not see that subtraction is useful for solving the problem.

Middle Concrete Operational Stage. This is a period of well-established concrete logic of differences, classes, and equivalences. Children at this stage do not see interrelationships or, at least, do not consider interrelationships in the data. Reversibility of operations is unavailable to them. They can use more operations and larger numbers, and more propositions in verbal problems, which indicates a raised level of abstraction. Nevertheless, they demonstrate that their

[1]The stages are adapted from Piaget by Collis (1975b).

Developmental Base Stage with Minimal Age	SOLO Description	1 Capacity	2 Relating Operation	3 Consistency and Closure	4 Possible Response Structure
Formal Operations (16+ years)	Extended Abstract	*Maximal:* Cue + relevant data + inter-relations + hypotheses	Deduction and induction. Can generalize to situations not experienced	Inconsistencies resolved. No felt need to give closed decisions—conclusions held open, or qualified to allow logically possible alternatives. (R_1, R_2 or R_3)	Cue x x x Response R_1 R_2 R_3
Concrete Generalization (13–15 years)	Relational	*High:* cue + relevant data + inter-relations	Induction. Can generalize within given or experienced context using related aspects	No inconsistency within the given system, but since closure is unique so inconsistencies may occur when he goes outside the system	x x x R o o o
Middle Concrete (10–12 years)	Multistructural	*Medium:* cue + isolated relevant data	Can "generalize" only in terms of a few limited and independent aspects	Although has a feeling for consistency can be inconsistent because closes too soon on basis of isolated fixations on data, and so can come to different conclusions with same data	x x x R o o o

Early Concrete (7–9 years)	Unistructural	*Low:* cue + one relevant datum	Can "generalize" only in terms of one aspect	No felt need for consistency, thus, closes too quickly: jumps to conclusions on one aspect, and so can be very inconsistent	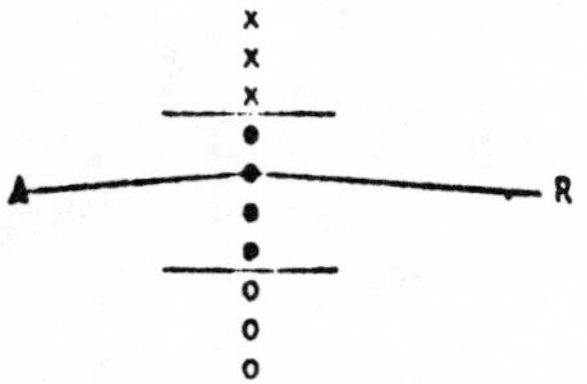
Pre-operational (4–6 years)	Prestructural	*Minimal:* cue and response confused	Denial, tautology, transduction. Bound to specifics	No felt need for consistency. Closes without even seeing the problem	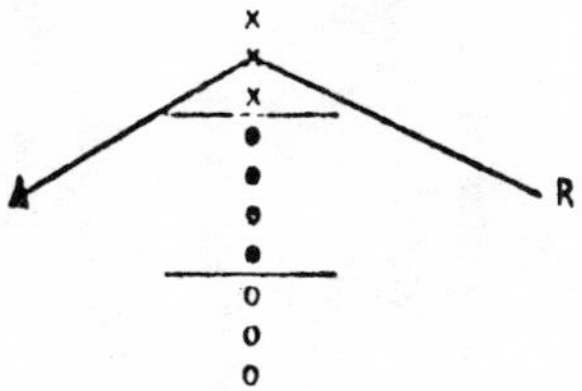

KEY: Kinds of data used: x = Irrelevant or inappropriate = Related and given in display = Related and hypothetical, not given.

FIG. 12.1. Base stage of cognitive development and response description. (From *Classroom examples of cognitive development phenomena: The SOLO taxonomy*. By K. F. Collis and J. B. Biggs, Hobart. The University of Tasmania, 1979.)

reasoning is clearly confined to the physical world of the present. The equivalent SOLO-level response, *multistructural,* indicates that the child comes to a conclusion on the basis of a sequence of discrete pieces of information *selected* from the data. In mathematics, the child can effectively use sequences of closures if the numbers are small, although if large numbers are involved fewer meaningful relations or operations can be handled. An overall view of the interrelationships between the operations and elements of a statement seems to be lacking. It is as if the statement represents a series of instructions to be performed in sequence.

Concrete Generalization Stage. This is the high point of concrete operational logic, its most significant feature perhaps being an ability to generalize from several concrete instances. No abstract hypotheses are considered and thus generalizations are often inadequate. The student often indicates an ability to interrelate given data but not to go outside it. The equivalent SOLO level is called *relational.* At this level the child's response indicates the capacity to relate one part of a system to another in a quite concrete way. A child can keep track of key interrelationships within a given numerical statement.

Formal Operational Stage. This stage is clearly associated with mature formal logic. The adolescent at this level is able to take all the data and interrelationships into account after obtaining an overview of the problem and considering an appropriate abstract hypothesis or generalization. This hypothesis or generalization is tested against the given data and against other considerations that may not be explicitly stated but are germane to the problem. The equivalent SOLO level involved is the *extended abstract* response. This level of response reveals an ability to deal with complete abstractions so long as they are within a well-defined system. This last is the hallmark of this level of response. Variables are no problem and do not need reference to physical analogues; mathematical and logical operations take on a reality of their own; there is an ability to view balanced systems such as equations as a whole and to work with them ensuring that the system remains in balance.

In summary, the relationship between cognitive development and SOLO, and the general characteristics of the latter, are outlined in Fig. 12.1. At the extreme left is the developmental stage. Next follows the name we have given to the SOLO level of response, and the columns marked 1 to 4 describe the characteristics of each SOLO level.

Capacity refers to the availability of working memory that the different levels of SOLO require. Functional working memory capacity increases with age, as does the space required for higher-level responses.

Relating operation refers to the relationship between the stimulus and certain aspects of the response. In the case of the prestructural response there is no logical relation; the cue and response are tangled into one unit.

Consistency and closure refers to two opposing needs felt by the learner: one is the need to come to a conclusion of some kind (to close); the other is to make consistent conclusions so that there is no contradiction either between the conclusion and the data, or between different possible conclusions. The greater the felt need to come to a quick decision the less information will be utilized, so that the probability that the outcome will be inconsistent with the original stimulus, the data, or other outcomes, is increased. On the other hand, a high level of need for consistency ensures the utilization of more information in making a decision, so that the decision is likely to be more open.

Possible response structure is an attempt to represent these characteristics in diagrammatic form. The student may respond to the stimulus by using three types of information: (1) irrelevant information (represented by x); (2) information that is contained in the original display, i.e., lesson, prior information, etc. (represented by ●); and (3) information and principles that are not given but are relevant, hypothetical, and often implicit (represented by ○).

THE RESPONSE MODEL IN RELATIONSHIP TO ADDITION AND SUBTRACTION

Direct application of the response model in mathematics is difficult because the actual response made by the child does not usually indicate the complexity of the thought process that gave rise to it. For example, if a child is asked to find the value of "x" in "$x + 3 = 9$," and responds correctly with the answer "6," one does not know whether this has been achieved by the simple unistructural device of counting on or by an extended abstract approach that would involve general principles such as inverse operations, associativity, and identity elements. To find the complexity of the process used by the child further probing is necessary. The illustrative examples from mathematics that follow assume this further probing.

Prestructural Responses. A response at the prestructural level shows that the respondent is not employing thought structures that would enable handling numbers and their mathematical combinations meaningfully. That is not to say that a child responding at the prestructural level could not answer the question: "What is 2 plus 3?" However, being able to respond "5" does not necessarily mean that the individual possesses the knowledge or the experiences to be sure that any group of 2 elements combined with a further group of 3 elements always gives a group of 5 elements. Being able to respond correctly to the task of matching a set of numerals with a set of 5 objects to be counted need not imply that the child understands the number idea "five," which incorporates the notions of invariance and conservation. For instance the child might be able to use the counting sequence accurately up to a certain point but still give nonconserving responses to some of the classical number conservation tasks.

The prestructural responses indicate that the individuals concerned are developing bases for operational thinking, but not until the next response level are the necessary thought structures available to deal adequately with number concepts, groupings, pairing corresponding elements, and so on.

Unistructural Responses. Responses at this level show that the individual is operating on elements based in immediately observable physical experience (e.g., 7 but not 759). Both the elements and operations of ordinary arithmetic are related directly to empirical experience. Responses to items using either large numbers or several operations in sequence seem to indicate that the child does not find the question meaningful. The idea of "inverse" is physical; for example, subtraction is seen as "what is put down can be taken up." The typical unistructural response to the question of why $x = 4$ in the statement $x + 3 = 7$ is to use a counting-on procedure of some kind. If pressed further most respondents confining their answers to this level will deny that "$x = 7 - 3$" is a legitimate method of solving the problem.

In summary then, *unistructural responses* are marked by a *single direct relationship to concretely available criteria.*

Multistructural Responses. Responses in this category are marked by apparent ability to handle logical operations provided these operations can be applied directly to particular experiences. Operations are not related to one another nor to abstract systems that are independent of experience. Thus, in dealing with the four operations of arithmetic, the pupil responding at this level regards the result of each operation as unique and can cope well with a statement that involves a sequence of discrete closures as long as there is no necessity to keep track of relationships between the operations in the statement. The arithmetical operations seem to be seen as reflections of reality—the sums and differences refer to real sets of objects whether present or not.

At this level of response the pupil still tends to work with qualitative comparisons. Thinking depends on the individual's perception of what is real and is marked by an inability to set up an empirical system based on measurement. It appears that children at this level perceive consistency between their qualitative comparisons, which at times leads them to consider quantitative comparisons, but using additive procedures only. Thus in their responses ordering elements by using direct excess comparisons is fairly common but comparison by ratio is not.

Meaningful use of the four operations of arithmetic is displayed only if the uniqueness of the result is guaranteed by previous experience. The experience must have been obtained with the operations themselves as well as the elements operated on. If the elements are related to direct physical experience a number of elements may be handled. However, if the elements are more abstract, only a very limited number can be dealt with. Thus a pupil responding at this level will cope with items like $3 + 5 + 2$ because the elements are verifiable; the $3 + 5$ can be closed to yield another verifiable element, thus allowing closure with $8 + 2 = 10$. A second type of item that can be handled is $475 + 234$; here the operation is familiar and the process can be carried out because the pupil knows from previous experience that a unique result will always be found.

Closure is still the only guarantee of uniqueness and the ability to handle a series of operations with small numbers by a sequence of meaningful closures seems analogous to using a sequence of given propositions to support a particular conclusion in verbal content areas.

The concept of "inverse" that comes through at this level of response is that of a "destroying" process. Although, in the classroom, this kind of thinking is difficult to distinguish from thinking where the inverse is seen as an "undoing" process, perhaps the key to the student's notion of "destroying" is the *irreversible quality* implied. For example, the student regards subtracting as destroying the effect of previously adding the same number without specifically relating to the operations themselves. To find the value of "x" in "$x + 3 = 7$," "x" is regarded as a unique number to which "3" has been added; subtraction of "3" *happens* to destroy the effect of the original addition. Probing usually reveals that the child explains the correct response to the question, not by referring to the addition and subtraction operations, but by adopting a primitive strategy, for instance, counting or by saying the equivalent of, "That's how I was taught."

Relational Responses. Responses at this level seem to rely on the same basic skills as at the multistructural response level; the major advance is less reliance on seeing uniqueness in the results of operations. There is still, however, a requirement for uniqueness of outcome. This guarantee of closure is obtained by making a generalization from a number of particular concrete instances. The individual response is still bound to the given empirical data but the respondent now seems prepared to infer beyond what can be demonstrated by modeling, and to generalize from a number of specific cases. Thus the response often takes the form of a concrete generalization where a few specific positive instances guarantee the reliability of a rule. Typically the responses indicate that the student is looking for positive instances to form a generalization, but is unaware of the need to check for counterexamples and limiting conditions. Moreover, as the response is based on the immediately available empirical evidence only, the respondent takes no account of hypothetical instances.

The skill of being able to handle a statement involving a series of closures is generalized in two ways. First, the size of the numbers used becomes irrelevant; second, and more significantly, the child now indicates an ability to keep track of the relationships that exist *within* the given statement.

Relational responses show an ability to work with generalized (apparently abstract) elements based on a few specific verifiable instances (e.g., $2x + 3x =$

$5x$). Investigation reveals, however, that the uniqueness of the result (the requirement of closure at this level) must still be guaranteed. In the example just given, questioning reveals that the child's view of the correctness of the statement most likely rests on some concrete analog such as "2 apples together with 3 apples gives 5 apples" rather than on any abstract mathematical generalization such as the distributive law.

At this level of response the inverse process becomes an undoing of an operation previously performed. It clearly possesses a reversible quality, but is limited in that it can be applied only to familiar operations.

Individuals responding at this level appear able to use an abstract rule obtained by generalizing a number of specific instances. However, they do not demonstrate the cognitive structures necessary to use the rule in a slightly different situation. The necessity for a guarantee of closure seems to inhibit the ability to view the data as a whole and focus on abstract relationships between variables.

Responses at the relational level are given on the basis of concrete generalizations where a few specific instances satisfy the respondent of the reliability of a rule and where the result of an operation, even though it be on apparent variables, is necessarily considered unique. In other words, *the individual sees relationships between the elements and the operations of the immediately available concrete system and works with them competently*.

Extended Abstract Responses. Responses at this level show that students reason and consider ideas at the abstract level and do not rely on experiences in the physical world. Responses are based on deductive reasoning from carefully chosen hypotheses or premises. Abstract variables that may have a bearing on the solution to the problem are visualized and manipulated. Respondents are no longer satisfied that one or two positive instances are a sufficient basis from which to generalize; a comprehensive inductive procedure is involved. The student responding at this level looks upon closure and uniqueness as abstract properties that have certain implications if they are available. The ability to operate on operations is in evidence in contrast to the earlier levels where the responses relate mainly to operating on elements. This ability to work without the requirement of uniqueness of the elements allows satisfactory solutions to be made to items where the problem is to decide on the equality or inequality of pairs of expressions, such as: $a + b$ and $(a + 1) + (b + 1)$. The level of difficulty in items such as this is a function not only of the degree of abstraction of the elements but also of the structural properties of the operations themselves.

In summary, extended abstract responses have the following characteristics: acceptance of lack of closure, the use of the reciprocal operation, and the ability to work with multiple interacting and abstract systems. All these characteristics *involve a comprehensive use of the given information together with related hypothetical constructs*.

CONCLUSION

Focusing the educator's attention on response levels and their structures has major implications for the classroom educational process. First it changes the emphasis from the level of cognitive development and/or IQ, about which the individual teacher can do very little, to the structure of the child's response, which enables a dialogue to begin with a view to raising the child's level of responding.

The response model also gives the teacher a rational basis for evaluation of the child's level of functioning. It indicates clearly that in mathematics "the correct answer" can be obtained in a number of ways and is not in itself sufficient to tell the teacher how well the child understands what he or she is supposed to be doing.

Much work needs to be done in mathematics education before the full benefits of focusing attention on the structure of these responses can be realized. One field where the concept can be put to use with good effect would be curriculum development. Normative studies indicating the expectations that the teacher might have of a normal age/grade group in terms of response complexity should benefit schools and teachers involved in school-based curriculum development.

Finally, the notions explicated suggest that we investigate closely the initial understanding of addition and subtraction with which the child begins school. Children do a great deal of learning before they come to school and indeed develop quite sound problem-solving skills appropriate to their own level of cognitive functioning. It seems illogical to ignore this and begin by imposing a potentially more efficient adult model on the children—a model that the children have not the cognitive capacity to come to terms with.

Children might well see the structure of the following problem as a "take-away": "Mary has 8 balloons and she gives 5 to Mark. How many has she left?" On the other hand, children might view the following problem as some kind of "matching": "Mary has 8 balloons, Mark has 5. How many more balloons has Mary than Mark?" For the teacher to categorize these both as mere applications of the subtraction algorithm could well be one of the first steps towards teaching children that mathematics is not something *one does* or *thinks with* but only a sequence of unconnected school-valued skills that supposedly will be of use later.

REFERENCES

Biggs, J. B. Developmental processes and learning outcomes. In S. Modgil & C. Modgil (Eds.), *Toward a theory of psychological development*. Slough, Bucks: NFER, 1980.

Collis, K. F. Concrete operational and formal operational thinking in mathematics. *The Australian Mathematics Teacher*, 1969, *25*(3), 77–84.

Collis, K. F. A study of concrete and formal reasoning in school mathematics. *Australian Journal of Psychology,* 1971, *23*(3), 289–296.

Collis, K. F. A study of children's ability to work with elementary mathematical systems. *Australian Journal of Psychology,* 1973, *25*(2), 121–130.

Collis, K. F. *Cognitive development and mathematics learning.* London: Chelsea College, University of London, 1974.

Collis, K. F. The development of formal reasoning. (Research report) University of Newcastle (N.S.W.), 1975. (a)

Collis, K. F. *A study of concrete and formal operations in school mathematics: A Piagetian viewpoint.* Melbourne: Australian Council for Educational Research, 1975. (b)

Collis, K. F. Levels of cognitive functioning and selected curriculum areas. In J. R. Kirby & J. B. Biggs (Eds.), *Cognition: Development and instruction.* New York: Academic Press, 1980. (a)

Collis, K. F. School mathematics and stages of development. In S. Modgil & C. Modgil (Eds.), *Toward a theory of psychological development.* Slough, Bucks: NFER, 1980. (b)

*Collis, K. F., & Biggs, J. B. *Classroom examples of cognitive development phenomena: The SOLO taxonomy.* Hobart: The University of Tasmania, 1979.

*Note: This research report is now out of print but forms the basic ms. for the following text Biggs, J. B., & Collis, K. F. *Evaluating the Quality of Learning: The SOLO Taxonomy.* N.Y. Academic Press Inc., 1981.

13 Type 1 Theories and Type 2 Theories in Relationship to Mathematical Learning

Richard R. Skemp
University of Warwick

I believe we need to make a global distinction between two categories of theory that I shall call Type 1 and Type 2, and that if this distinction is not made, there is likelihood of methodological errors, not only of detail but of principle.

My own realization of this distinction has followed the construction of a new model of intelligence, offered as an alternative to the psychometric group of models that have dominated the field for 70 years. In the course of discussions about research based on this model, and particularly when discussing how it could be tested, it became apparent that an inference from this model was that the methods by which it should be tested were not necessarily those traditionally used by experimental psychologists.

In this regard let me stress the relationship between theory and method. Methodology refers to the set of techniques by which a researcher builds and tests a theory. This includes both constructing a new theory ab initio, and improving an existing theory by extending its domain or increasing its accuracy and completeness. Methodology and theory are thus closely related, and the construction of a successful theory will depend largely on the use of appropriate methodology.

A NEW MODEL OF INTELLIGENCE

As a starting point I present a brief account of this new model. (A full exposition is to be found in Skemp, 1979.) The model assumes, as a matter of observation, that much, possibly most of human behavior is goal directed, which implies that if we want to understand adequately what people are doing, we need to go beyond the outward and easily observable aspect of their actions and ask ourselves what is their goal. To help in thinking about how people direct their actions in ways that lead to the achievement of their goals, a model of a *director system* was developed, which synthesizes ideas from cybernetics and cognitive psychology. Its essential features include: (1) some kind of sensor that takes in the present state of the operand (the operand being whatever is to be taken from the present state to the goal state); (2) some kind of internal representation of the goal state; (3) a comparator, which compares these two; and (4) a plan, by which energy is applied to the operand in such a way as to diminish the difference between its present state and its goal state until these coincide.

The changes from present state to goal state take the operand through a succession of intermediate states, each of which becomes a present state. All of these, in turn, have to be represented within the system so that they can be compared with the goal state. It is a short step from this to the need for a mental representation of the path from the present state to the goal state. This is a minimal requirement. More effective, particularly in varying environments, is to have not just an image of a particular path, but a cognitive map from which a variety of paths can be constructed, as required, to meet the requirements of different starting points and environmental conditions. A schema, or conceptual structure, is simply a further development of the idea of a cognitive map, including concepts at different levels of abstraction and a symbol system for retrieving and manipulating these concepts.

In the lower animals, many of these director systems are innate, the result of natural selection. But there is an upper limit to what can be transmitted genetically; there are other disadvantages, such as slowness to adapt to environmental change. So it is not surprising that some species have evolved the ability to set up new director systems and to improve the ones they have. This is how *learning* is conceptualized within the present model. Other animals can learn too, but we have also evolved a more advanced kind of learning that is qualitatively different from those studied in animal laboratories and embodied in theories such as operant conditioning. It is the ability to learn in this more advanced kind of way that I now call *intelligence*. A major feature of intelligence is the construction (building and testing) of the schemata (conceptual structures) that are considered to be an important part of more advanced kinds of director systems.

The new model uses the concept of a director system at two levels. (Fig. 13.1). Leaving out all the interior detail, delta-one (Δ_1) is a director system whose operands are physical objects in the outside environment. Delta-two (Δ_2) is a second-order director system, which has delta-one (Δ_1) as its operand. Its function is to take delta-one to states in which delta-one itself can function better. This includes not only improving director systems, but bringing new ones into existence. In brief, *delta-two optimizes delta-one*. Learning is one of the long-term ways in which this is done; making particular plans for particular situations is another, short-term; building up a stock of plans to keep on hand for regularly encountered situations is another. Algorithms are an example of the last.

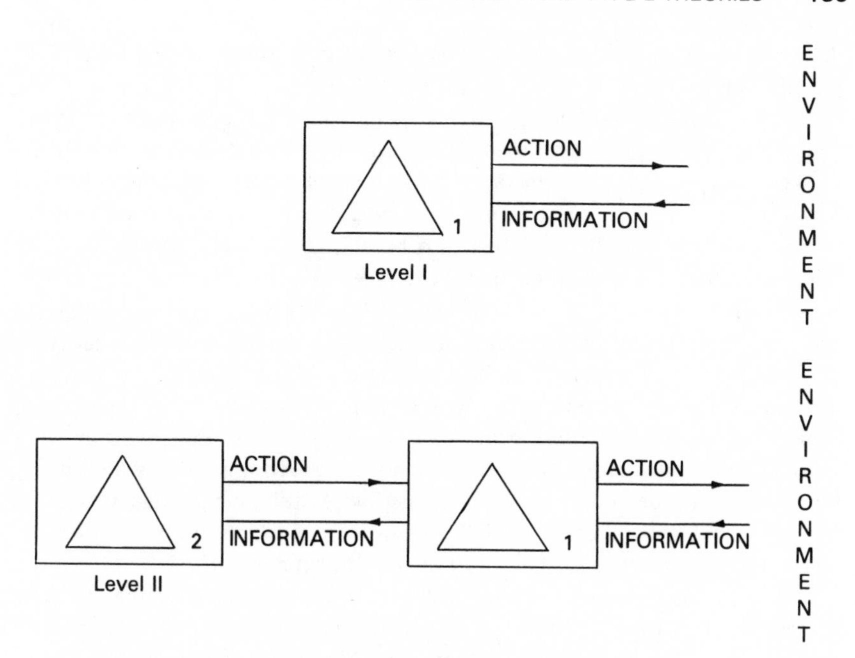

FIG. 13.1. The concept of a director system (Δ) at two levels.

With the help of the foregoing, we can now distinguish two major categories of theories.

A *Type 1 theory* is a somewhat abstract and general mental model of regularities in the physical world. It is thus a particular kind of schema (for we can imaginatively construct other kinds of schemata that do not and are not intended to represent anything having physical existence). A Type 1 theory is used by delta-one as a basis for goal-directed action on operands in the physical environment. A theory is a cohesive and abstract body of knowledge (knowledge-that) from which we can, as required, derive particular procedures (knowledge-how) to achieve particular goals in particular situations. Knowledge-how is a particularly important case of prediction. A prediction states that initial state *A*, without intervention, is followed by state *B*. Example: astronomical theory. Knowledge-how takes this a step further, and states that initial state *A*, with intervention based on plan *P*, will result in state *B*. Example: theory of electronics. Knowledge-how is a necessary but not sufficient condition for being able to do something. The intervention prescribed by the knowledge-how may be beyond our ability to translate into action.

All the natural sciences such as chemistry, metallurgy, electromagnetic theory, or genetics, are Type 1 theories. In their respective fields of application, they are very successful in helping us to direct our actions successfully at the delta-one level, which is to say in achieving goal states of operands in the physical environment.

A *Type 2 theory* is a model of regularities in the ways by which Type 1 theories are constructed, and by which plans of action (for execution by delta-one) are derived from these theories. It is a mental model of the mental-model-building process. From an appropriate Type 2 theory, we may hope to derive knowledge–how; possibly we shall also be able to intervene helpfully in children's learning of mathematics. Some examples of Type 2 theories are constructivism (Steffe, Richards, & von Glaserfeld, 1979); my own theory of intelligence; and any theory about the learning and teaching of mathematics that recognizes that teaching is an intervention in someone else's learning (i.e., that regards learning as a goal-directed activity with an important degree of autonomy in the subject, rather than regarding behavior as being shaped by the environment).

A *Type 1 methodology* is concerned with constructing (building and testing) the models that delta-one requires for its successful functioning. When constructed, these models are Type 1 theories. Each of the natural sciences has its own methodology, though these have much in common.

A *Type 2 methodology* is concerned with constructing (building and testing) models of how Type 1 theories are constructed, and how particular plans of action are derived from these. When constructed, these models are Type 2 theories.

The importance of the foregoing is that if Type 1 and Type 2 theories belong to different categories, then we must be very alert to the possibility that they require different methodologies. Failure to make this distinction may result in the application of inappropriate methodologies, leading to unsound theories. In addition, it may result in the wrong overall conception of what one is trying to construct, so that while working on a Type 2 theory a person is all the time trying to make it look like a Type 1 theory.

To help show how a Type 2 methodology needs to differ from a Type 1 if it is to succeed, the following summary of Type 1 methodology is offered as a starting point. (Modified from Skemp 1979, p. 163, Table 7.)

Building	*Construction Mode*	*Testing*
One's own experience of the physical world	1	*One's own experiments* on physical objects, involving the testing of predictions.
Communications from others; personal, lectures, journals; searching the literature.	2	*Comparing* one's own ideas with those of others, often involving *discussion;* seminars, conferences.

Building	*Construction Mode*	*Testing*
From within, by working on and with existing ideas: synthesis, extrapolation, imagination, intuition. *Creativity*.	3	Comparison with one's own existing knowledge and beliefs: *internal consistency*.

Although a correspondence can be seen between the three kinds of building and of testing, any one or more of the former can be used in conjunction with any one or more of the latter in the construction of a theory. The natural sciences use all three modes of building and all three modes of testing. However, the ultimate appeal is always to testing by Mode 1, experiment. If the purpose of constructing (which includes improving) Type 1 theories is to increase the powers of delta-one relative to the physical world, the physical world is where they must prove their success. Other criteria, such as economy, coherence, intelligibility, are also important. They help to make a theory more usable, by facilitating the conversion of knowledge–that into knowledge–how.

Popper (1976) proposes that the term "scientific" should be reserved for theories tested by Mode 1. This would be to equate all sciences with the natural sciences, and any scientific theory would thus be a Type 1 theory. Scientific method would in this case be a body of particular methods derived from the methodology summarized in the foregoing, with Mode 1 testing as an essential component. I do not yet know whether I myself accept this position.

What kind of theory is mathematics? We need at least a partial answer to this, for how can we usefully think about teaching it if we do not know what kind of a theory it is that we are trying to teach? Mathematics seems to me to be a Type 1 theory of an unusual, perhaps unique, kind. Though it can make good use of Mode 1 at the outset (e.g., in the building of the concept of order and in the initial construction of the natural numbers), it rapidly abandons Mode 1 and relies entirely on Modes 2 and 3. Thus, correct or incorrect predictions of physical events play no part in confirming or refuting a mathematical theory, as they do a major part for other Type 1 theories. But the discovery of an internal inconsistency would refute a mathematical theory. The discovery that new ideas were consistent with the accepted body of mathematical knowledge would help to confirm them, and a demonstration that they were a necessary consequence of certain parts of this knowledge would constitute a proof, in the mathematical sense.

Although mathematics is not itself one of the natural sciences, it can be regarded as a conceptual "kit" of great generality and versatility, so valuable to anyone who wants to construct a scientific theory as to be almost indispensable. The conversion from a mathematical statement to a theoretical model is often a very short one, requiring only the attachment of units. E.g., $E = IR$ is a mathematical statement if E, I, R represent pure numbers. But if they represent numbers of units of emf., current, and resistance respectively, it becomes Ohm's law. These very close links, and the ease of transition both ways, suggest that mathematics may be regarded as a Type $1x$ theory, having all the characteristics of a Type 1 theory except Mode 1 testing. Note that Mode 1 building may be present, as in the construction of the natural numbers.

A REFOCUSING

In order to write this chapter, it was first necessary to put a certain distance between myself and the ways of thinking I had acquired as a mathematician, with physics as a supporting subject. This was followed by a period of 18 years as a psychologist. My orientation was that of an experimental psychologist, but during this period, I now realize that I was engaged in making the transition from a Type 1 theorist to a Type 2 theorist. This is a transition that others have been making. But we who are making this transition are in a different position from persons working on Type 1 theories; though they are at the frontiers of knowledge, they have well-established methods of exploration. We are at two frontiers at the same time, the second one being a frontier of methodology. We need a methodology for investigation: not investigating children's observable performance, but whatever brings about changes in their ability to perform. These changes may result: (1) from increase of their knowledge–that, the construction and improvement of their mathematical schemata; (2) from their having now succeeded in deriving a new plan from their existing knowledge; (3) from increasing their repertoire of plans, eliminating for a greater number of tasks the necessity for (2).

These changes (in terms of the present model) take place within the child's delta-one, which by its nature cannot be observed by the experimenter. And whatever brings about these changes (in the present model, it is the higher-order system delta-two) is even more inaccessible to observation. By the activity of reflective intelligence, delta-two can sometimes observe, and even report on, activities within delta-one. But the activities of delta-two itself can only be inferred from changes in delta-one and the circumstances leading to such changes.

However, if we could find some way of observing the concepts and schemata within a child's delta-one, even though this would necessarily be indirectly and by inference, we would have made a substantial beginning. These observations, both for building and testing our theory-in-the-making, would then replace the Mode 1 methods described previously.

Our starting point toward a method is a consideration of the function of symbols. These act as an interface of two kinds. The first is between the child's mathematical schemata[1] (located in the child's delta-one), and the experimenter's. The second is between the conscious and unconscious levels of the child's own thinking. As I have suggested elsewhere (Skemp, 1979, pp. 157–158) it is questionable whether secondary concepts and schemata can be observed directly, even by their possessor; our sense organs are directed outwards, towards the physical world. The process of making a concept conscious seems to be closely connected with associating the concept with a symbol. So it is by symbols that children know what is in their own mind, and it is by symbols that the experimenter is enabled to know what is in the child's mind. This knowledge is only partial, but it is the best we can get.

From the foregoing analysis, diagnostic interviews and teaching experiments both emerge as methods appropriate for the construction of a Type 2 theory. In a diagnostic interview, the experimenter sets up internal, tentative images of what might be in the child's delta-one, and tests these by symbolic interactions with the child. In other words, the experimenter tries to get inside the child's mind by imagining the child's mind—by constructing a mental picture of its workings—and interpreting what the child says in relation to this image. This interpretation is tested by another question, and if necessary, is corrected. In this way, the experimenter may hope to construct images of the thinking of a number of children, and in particular of the ways in which they construct plans from their available schemata. From these, regularities may possibly be abstracted and put together into a theory.

The method of the teaching experiment takes this process a step further. The present state of the mathematical schemata of the learner is modeled in the mind of the experimenter, who also decides on a goal state, the object of the teaching procedure. The experimenter can conceptualize this; the learner cannot, or can do so only vaguely. Next, the experimenter makes a conceptual analysis of the concepts belonging to the goal schema and reanalyzes these in turn. This yields a dependency network to show which concepts are prerequisite for others, or at least a working hypothesis about these relationships. On the basis of this, the experimenter sets up a path connecting the starting schema with the goal schema. This operation takes place in the experimenter's own delta-one, and the path will be a psychological path, not a logical one. That is to say, it will be a sequence of schemata, each of which can be reached from the one before by the expansion of existing concepts, the formation of new concepts, or extrapolation. It will not be a logical sequence, which is one of logical inference. The latter involves the examination of implications between concepts that a person already has; it is not a process by which new concepts can be formed. As Steffe *et al.* (1979) argue: "Concepts, structures, skills, or anything that is considered knowledge cannot be

[1] *Schema*. A conceptual structure. Can be derived from the idea of a cognitive map, if we regard a schema as analogous to a cognitive *atlas* in which (e.g.) a dot representing London or New York on a map of U.K. or U.S.A. can itself be expanded into a map. Not quite the same as Piaget's "scheme."

conveyed ready-made from teacher to student or from sender to receiver. They have to be built up, piece by piece, out of elements which must be available to the subject [p. 43]." The teaching experiment may be regarded as an extension of the diagnostic interview, in which the purpose is to make and test hypotheses not only about the nature of a child's thinking at a particular time, but about how this thinking is developed from one stage to another.

The teaching experiment itself will involve trying to take the learner along the path from starting schema to goal schema by two means: presentation of the material that has been devised and additionally, where necessary, by explanations and direct information that help the learner to assimilate the new material to a currently available schema. The method of the diagnostic interview will be used at every stage to compare the state (of this schema) that the learner should have reached, as imaged on the path within the teacher's delta-one, and the state at which the learner has in fact arrived. In this way, the experimenter will try to correct the teaching plan as initially devised, until one is developed that achieves the desired learning goal, so far as can be determined from the diagnostic interviews. It will then be necessary to discover whether these plans are effective for other teachers and learners. This is the field in which a Type 2 theory for the teaching of mathematics will have to prove itself, corresponding to the proving ground in the physical world of a Type 1 theory.

The foregoing combination of methods appears to me an appropriate replacement for Mode 1 building and testing, as a first step in the conversion of the Type 1 methodology to a Type 2 methodology. Modes 2 and 3 do not need replacing, although their relationship may need to be reconsidered. These methods are already in use, having been devised quite independently of the new model of intelligence that has been the starting point for the foregoing analysis. This convergence of thinking I find encouraging.

REFERENCES

Popper, K. *Unended quest: An intellectual autobiography,* Glasgow: Fontana/Collins, 1976.

Skemp, R. R. *Intelligence, learning, and action*. Chichester: Eng. John Wiley & Sons, Inc. 1979.

Steffe, L. P., Richards, J., & von Glaserfeld, E. Experimental models for the child's acquisition of counting and of addition and subtraction. In W. Geeslin (Ed.), *Explorations in the modeling of the learning of mathematics*. Columbus, Ohio: ERIC/SMEAC, 1979.

14 The Development of Addition in the Contexts of Culture, Social Class, and Race

Herbert P. Ginsburg
University of Rochester

This chapter is concerned with the influences of culture, race, and social class on the development of addition ability and on other aspects of mathematical cognition.

Studies described in this chapter were based on a conceptual framework describing the development of mathematical thought (Ginsburg, 1977). According to this view, knowledge of elementary mathematics may be conceptualized in terms of three cognitive systems operating concurrently in individual children.

System 1. Before entering school, or outside the context of formal education, children develop techniques for solving quantitative problems, including those that do not demand a numerical response. For example, consider the *perception of more* originally described by Binet (1969). Given two randomly arrayed collections of dots, a child of 3 or 4 years of age can easily determine which of the two collections has "more" than the other, at least when relatively small numbers of elements are involved. The child's speed of judgment (only a second or 2 is needed) indicates that counting is not necessary to solve the task. Other examples of System 1 skills involve one-to-one correspondence, equivalence, and seriation, as described by Piaget. Because System 1 develops outside the formal school setting, it may be termed *informal*. Further, because System 1 does not involve counting or other specific information or techniques transmitted by culture, it may be termed natural.

System 2. During the preschool years, the child begins to say the counting numbers and to count objects. Gradually, the range of counting activities is expanded: The child learns to say larger numbers; to count greater numbers of objects; to use counting to solve various practical arithmetic problems; and eventually to perform mental addition. System 2 (of which counting is a major component) is *informal* insofar as it develops outside the context of schooling, but it is *cultural* because it depends on social transmission through adults, television, books, and the like.

System 3. The child attending school is taught symbolic, codified arithmetic and other forms of elementary mathematics. The child encounters written symbolism, algorithms, and explicitly stated mathematical principles. These cultural inventions and discoveries are more powerful than counting and, used properly, can provide the child with considerable efficiency in problem solving. The paradox is that so many children, well endowed with informal mathematics, fail to make effective use of the even more powerful cultural legacy. Because System 3 typically develops in school or through contact with written culture, it may be termed *formal.* System 3 is obviously transmitted by social agents and hence may be termed *cultural*.

Research shows that certain System 1 skills, particularly those studied by Piaget, are extremely widespread and may be found in members of diverse cultural groups (Dasen, 1977). Is the same true of other System 1 skills like the *perception of more?* Do they emerge naturally and spontaneously in lower-class children, in blacks, in members of traditional cultures? Perhaps System 1 skills are as universal as basic language ability. If so, they might serve as a solid foundation for mathematical education.

As we have seen, System 2 skills like *mental calculation* develop in the absence of formal schooling but are transmitted through cultural agents. We know from anthropological accounts like Zaslavsky's (1973) that certain illiterate cultures, but not others, develop calculation skills, usually based on counting. But how powerful are these skills? How widespread are they across cultures and races? And how is their development affected by formal education?

System 3 skills like *written addition* are obviously the products of cultural transmission, especially schooling. Yet despite the importance of such academic knowledge, we know little about it. How is written addition organized in the minds of those who learn it? Do children from different cultures organize it in different ways? How is written addition related to the child's already existing system of informal knowledge? Questions of this type are obviously basic for educational practice.

The three studies described in the following sections attempt to answer some of these questions. Our first study investigates the effects of schooling and culture on the development of a System 2 skill, mental addition. The second study investigates the nature of a System 3 skill, written arithmetic. The aim is to determine how written arithmetic develops in children of different cultures, African and American, and how it is affected by already existing systems of informal mathematical thought (Systems 1 and 2). The third study investigates the effects of race and social class on the development of American children's mathematical thought. The focus is on a System 1 skill, the *perception of more*, and on a System 2 skill, *addition calculation*.

STUDY 1: MENTAL ADDITION IN CROSS-CULTURAL CONTEXT

Study 1 is concerned with the effects of culture and schooling on the development of mental addition. Hebbeler (1977) has shown that at the ages of 3 and 4, children perform reasonably well on addition problems involving *concrete objects*. When asked to add three candies and two candies, all actually present, young children generally use some kind of counting strategy to achieve success. They sometimes count the members of both sets, sometimes count on from the larger number, sometimes count with fingers, and sometimes count in their heads. Although successful with concrete objects, young children experience difficulty when asked to add imaginary objects or spoken numbers. At this age level, such mental addition appears to be beyond their capacity. Around the age of 5, children begin to develop effective procedures for mental addition. Over a period of several years, they progress from the use of informal counting strategies to learned number facts and then to "invented procedures" (Houlihan & Ginsburg, 1981) that may exploit what has been learned from adults or teachers but are at least in part the child's own invention (Groen & Resnick, 1977).

How does mental addition develop in other cultures, particularly illiterate ones? Zaslavsky (1973) presents historical and anthropological evidence showing that unschooled, illiterate Africans are capable of impressive achievements in mathematical work. Such anthropological and historical findings need to be supplemented with psychological research. Hence, the first aim of Study 1 (Ginsburg, Posner, & Russell, 1981b) was to examine the development of informal addition in unschooled Africans from childhood to adulthood, and to obtain comparable data from American subjects.

A second aim of the study was to investigate the effect of schooling on mental arithmetic. In general, most writers have maintained that Western schooling has beneficial effects on cognitive processes. For example, Scribner and Cole (1973) propose that schooling facilitates the generalization of solution rules and the verbalization of cognitive processes. Yet the evidence in this area is contradictory. Some studies (Stevenson, Parker, Wilkinson, Bonnevaux, & Gonzalez, 1978) find major effects of schooling on a variety of cognitive processes, including memory and reasoning. Other investigations (Sharp, Cole, & Lave, 1979) find some areas in which schooling has positive effects, and some where it seems to make no difference. Still other investigations (Kiminyo, 1977) find no important effects of schooling on Piagetian tasks.

We were interested, in particular, in the effects of schooling on the development of informal mathematical thinking. Posner (in press) found that schooling does not affect Dioula children's performance on elementary addition tasks involving concrete objects; both schooled and unschooled Dioula children performed at a similar high level. However, the addition problems employed were extremely simple, involving single-digit numbers, so that a ceiling effect may have operated. That schooling has a beneficial effect on mathematical problem solving is suggested by Lave (1977) who found that among Liberian tailors schooling facilitated the solution of spoken arithmetic problems that were "similar to school book problems [p. 3]." Thus, there is a small amount of evidence that in at least some cultures school learning, presumably involving specific algorithms, can transfer to a degree to arithmetic problems that are not given in written form.

Study 1 involved a comparison of American children and adults with subjects of comparable age from the Dioula, a group in the Ivory Coast, a small country in West Africa. Over the past 20 years or so, the Ivory Coast has undergone a program of intense economic development with the assistance of the French, its former colonial masters. Several cities in the Ivory Coast, especially the capitol, Abidjan, are extremely modern. But, despite recent attempts at modernization, much of the country, particularly the rural areas of the north, remains relatively untouched by government influence.

The Dioula, a Moslem group, have traditionally engaged in mercantile activities and are scattered throughout West Africa. They speak a distinct language. Although some Dioula do engage in farming, in general Dioula society is commercially oriented. Many Dioula are now settled either temporarily or permanently in urban centers.

Some of our subjects were drawn from traditional Dioula villages. Others came from the city of Bouaké, a small community in the center of the Ivory Coast with a population of about 160,000. Most of our subjects from Bouaké can be considered newly urbanized individuals, whose families still maintain close links to the village culture.

Schools in the Ivory Coast have recently been transformed from rigid traditional environments to more progressive places of learning. There is increasing emphasis on active learning and children's manipulation of objects, many of which are taken from the Ivorian setting. Instruction is conducted in French, which can be considered a foreign language for most of the children as they begin school. The mathematics curriculum was developed by teams of European educators and psychologists and follows the French pattern in great detail. After the children view a particular mathematics lesson on television, the teacher uses a manual to expand on various ideas introduced. A heavy emphasis is placed on drill. The curriculum has both a "new math" flavor involving set theory notions and notation, and a Piagetian component, involving emphasis on manipulation of real objects and the gradual development of abstraction. The children's high degree of motivation and the intensity of their work under various difficult and uncomfortable physical conditions is impressive.

Subjects

The American subjects were 16 second-grade students (eight boys and eight girls, mean age 8.1 years) and 16 fifth-grade students (eight boys and eight girls, mean age 11.2 years) from a public school in Ithaca, New York. The American adults were 16 college students at the University of Maryland. These students were mainly freshmen and sophomores, eight men and eight women.

In the sample of schooled Dioula subjects, there were 16 children (eight boys and eight girls, mean age 9.8 years) at the third-grade level, and 16 children (eight boys and eight girls, mean age 12.6 years) at the sixth-grade level. Because the Ivorian schools in which we worked lacked a kindergarten, subjects were chosen from one grade in advance of their American counterparts so as to equalize years of education. The students were selected randomly at two different schools in the city of Bouaké. Schooled Dioula adults were sampled from two schools in Bouaké. Twelve subjects attended a private secondary school, conducted in the evening. The mean number of years of school for this group was 11.4. Their average age (self-reported) was 19. The remaining four subjects were students in a prestigious teacher training institute at Bouaké. The students averaged 14.7 years of school and had a mean age of 21.5.

Among the unschooled subjects, there were 16 9- to 10-year-olds (eight boys and eight girls); 16 12- to 13-year-olds (eight boys and eight girls) and 16 adults, all male as females refused to participate. The ages of the children are approximate, as few knew their birth dates. The adult subjects were solicited in the Bouaké market. All the subjects were Moslem, and none was literate, even though some had taught themselves to write numbers, and one subject could do written operations. The mean age of the merchants (self-reported and approximate) was 34.

Method

Eight mental addition problems, administered verbally, were given to each subject. The problems were in two sets each representing a different order of sum magnitude. One of the sets received by each age group was the same as that received by at least one other group. For instance, the youngest group received four problems where the sum was between 10 and 20, and four others between 20 and 40. The middle group was given the latter four problems and four more, whose range of magnitude was between 40 and 100. The problems for adults included the four large problems given the middle subjects and four more whose sum exceeded 700.

For African subjects, the problems were given in the preferred language. Among the unschooled subjects, the preferred language was always Dioula. For the schooled children, the predominant language preferred was Dioula; among the schooled adults, French was always preferred. The African examiner, fluent in both French and Dioula, stressed that the problems could be solved in any way desired, including by means of written procedures. After subjects solved a given problem, they were asked to describe their solution procedures. Both correctness of response and type of strategy employed were coded. A reliability check yielded about 90% agreement between two independent coders.

Results

Subjects were given a score of from 0 to 4 points according to the accuracy of response on each of the eight problems. A subject who used an appropriate strategy and also gave a spontaneous correct answer received 4 points. If the initial answer was incorrect, but was subsequently corrected by subject, the subject received 3 points. The results, given in Table 14.1, show high levels of competence in schooled subjects, American and Dioula, at almost all age levels, and reasonable proficiency in young unschooled Dioula followed by extremely strong performance by unschooled Dioula adults. T-tests showed that at the youngest age level, schooled and unschooled Dioula scored at significantly lower levels than the Americans, and that at the middle age level, unschooled Dioula scored at significantly lower levels than Americans.

The proportion of various strategies used on all problems combined is shown for the different groups in Table 14.2. In the case of young children, Americans and Dioula show a similar pattern of usage except that the unschooled Dioula tend to rely on number facts more and counting less than their schooled counterparts. In general all groups use strategies appropriate to the nature of the particular problem. For example, all groups make most use of number facts on problems involving 10 + 7 and 20 + 10. Children in all groups do *not* use number facts on larger problems that do not involve simple multiples of 10, like 19 + 12 or 14 + 13. The most frequent use of counting is on problems involving relatively small sums like 9 + 6. The highest proportion of regrouping strategies (e.g., 14 + 13 is 10 + 10 + 4 + 3) is on relatively large problems like 15 + 20 or 14 + 13.

The use of strategies is much different among middle children. Counting virtually disappears. The schoolchildren, both American and Dioula, use the mental algorithm half of the time, with considerable accuracy. The unschooled

TABLE 14.1
Mean Accuracy Scores by Age and Cultural Group

	American	*Schooled Dioula*	*Unschooled Dioula*
Youngest	29.5	26.3	23.9
Middle	29.9	29.5	26.4
Adult	31.0	30.6	30.0

Note. Maximum possible score = 32.

TABLE 14.2
Proportion of Usage and Success of Various Strategies on all Problems by Age and Cultural Group

	Number Fact	*Counting*	*Regrouping*	*Other*
Youngest children				
American	.22(100)	.46(81)	.21(88)	.11(81)
Schooled Dioula	.17(100)	.47(60)	.26(79)	.10(42)
Unschooled Dioula	.36(96)	.21(59)	.27(62)	.16(24)
Middle children		*Algorithms*		
American	.11(100)	.52(75)	.34(95)	.02(50)
Schooled Dioula	.13(100)	.53(79)	.32(85)	.02(0)
Unschooled Dioula	.23(100)	0(0)	.70(60)	.07(33)
Adults		*Algorithms*		
American	.01(0)	.68(93)	.19(96)	.11(87)
Schooled Dioula	.11(100)	.63(90)	.26(73)	0(0)
Unschooled Dioula	.44(96)	.03(0)	.52(76)	.01(0)

Note. Percentage of successful solutions is indicated in parentheses.

Dioula do not use the mental algorithm at all, as would be expected. Rather, they use regrouping, employing it twice as frequently as the schooled children.

All groups use number facts almost exclusively on two problems, 20 + 10 and 60 + 30, and are completely accurate. When the schooled children use the mental algorithm, they are wrong about 25% of the time. When the schooled children use regrouping they are seldom wrong. By contrast, when the unschooled Dioula use regrouping they are frequently incorrect.

Finally, among the adult groups a pattern similar to middle children's is shown. The adult Americans and schooled adult Dioula relied even more heavily on the mental algorithm than did the middle school children, using it about two-thirds of the time. By contrast, the unschooled adult Dioula relied on regrouping for about half the problems and also solved nearly half the problems by using number facts, which seems to represent an increase over the proportion used by middle unschooled Dioula. Adults were generally correct, regardless of the strategy employed.

Discussion

One aim of the study was to describe the development of mental addition in both unschooled and schooled individuals. The overall competence of the unschooled Dioula is an example of the effectiveness of informal cognitive skills within a culture favoring them. The unschooled subjects seem to achieve this competence by the increasingly efficient and discriminating deployment of strategies for mental addition. The Dioula begin with rather elementary counting procedures, but soon switch to the extensive utilization of regrouping methods, a more efficient strategy particularly in the case of larger numbers. The unschooled, particularly in adulthood, make significant use of remembered number facts, which presumably they have acquired through everyday experience in calculation. At first, the Dioula do not employ the various strategies with great accuracy (e.g., regrouping at the young age level is not entirely accurate). But gradually, the Dioula become increasingly proficient in the use of the strategies, and by the middle age period the Dioula learn to discriminate among different types of problems, applying different strategies where appropriate. This is an example of increased economy and efficiency in the development of cognitive operations.

The finding of slight superiority of schooled children over unschooled is somewhat different from Posner's (in press) data showing no difference between schooled and unschooled Dioula children on concrete addition problems. Presumably, in the present instance, schooled children do better than their unschooled peers because the addition problems are more difficult and abstract than in Posner's original study.

With respect to strategies, schooled African and American children are quite similar, though both use different calculation procedures from unschooled individuals. The most dramatic developmental phenomenon, setting apart the schooled from the unschooled, is the schooled children's extensive use of the standard algorithm, deployed mentally, at the middle age level. Like their unschooled peers, schooled subjects discriminate among problems, using various strategies where appropriate. Again, this is an expression of the developmental trend toward increased economy and efficiency in the use of cognitive activities.

The second aim of this study was to examine the effect of schooling on mental addition. In this respect, the results are quite clear: To a large extent schooled individuals transfer to mental addition problems the standard algorithm acquired in school. In the present case, schooling does not operate to improve problem-solving skills in general; rather, it provides individuals with a specific method that is useful for particular types of problems. In future research on the effects of schooling, it may be useful to distinguish between specific and general effects of schooling, particularly as they interact with culture. Perhaps this distinction may resolve some of the confusion presently existing in that literature. Our finding concerning the specific effects of schooling is similar to that of Scribner and Cole (1978), who found that literacy in an unschooled group also had specific and limited effects.

STUDY II: WRITTEN ADDITION IN CROSS-CULTURAL CONTEXT

Study II focused on two Ivorian groups, the Baoulé and the Dioula. As we have already seen, the Dioula are a Moslem mercantile group whose culture stresses mathematical activities. By contrast, the Baoulé are an animist agricultural group, placing no particular emphasis on mathematics. Indeed, Guerry (1975) comments that for the Baoulé "counting, measuring . . . [are] . . . the white man's specialty [p. 31]." So far as we know from ethnographic reports like Guerry's, mathematics does not play a central role in Baoulé culture. Just the reverse is true of the Dioula, who must use calculation procedures in trading. The two groups provide a useful contrast in terms of hospitality to mathematical ideas and procedures.

In contrast to earlier views concerning primitive mentality (Cole & Scribner, 1974), recent cross-cultural research on cognitive skills in general and mathematical thinking in particular leads us to predict that, after only a few years of training, African children from certain traditional societies should assimilate school arithmetic in much the same manner as Americans. This proposition may be elaborated as follows:

1. At the outset, Baoulé and Dioula children may experience difficulty in learning arithmetic because it is presented in French, a language with which most of them are not proficient. Also formal schooling is a novel experience for members of these cultures. Consequently, in the first few years of school, American children may display a firmer grasp of basic arithmetic principles and procedures than these African children. But as language proficiency and familiarity with school increases, the gap between Americans and Africans should narrow.

2. Because American, Baoulé, and Dioula children possess similar cognitive processes and systems of informal mathematical knowledge, the three cultural groups should assimilate school mathematics in roughly the same manner.

(a) Americans and Africans should acquire the same basic mathematical concepts. The system of arithmetic should have the same essential meaning for all groups.

(b) American children do not simply apply standard algorithms to solve written addition problems. Instead they often use invented strategies (Ginsburg, 1977; Groen & Resnick, 1977). These are assimilations of written techniques into already existing informal procedures. We predict that the Baoulé and Dioula should be as capable as Americans of the creative activity of developing invented strategies.

(c) Research shows that American children make certain common and widespread errors in arithmetic (Ginsburg, 1977), for example, faulty addition procedures, misalignment of the numerals, failure to carry, and misremembering number facts. The errors seem to stem from specific cognitive processes. Thus, a child adds 22 + 2 and obtains the result 44 because he assimilates the problem into the previously induced rule "Add all the numbers on the top with all the numbers on the bottom." In this example, the difficulty is an overextension of a rule previously found to be valuable within a limited context. We maintain that because both African and American children employ similar cognitive processes in general and quantitative procedures in particular, both groups should produce similar kinds of errors.

3. In view of the difference between the Baoulé and Dioula groups' cultural emphasis on mathematics, we predict that the Baoulé should experience more difficulty than the Dioula in learning school arithmetic.

To obtain data relevant to these predictions, we conducted a study (Ginsburg, Posner, & Russell, 1981a) comparing the performance of American, Baoulé, and Dioula children, at two age levels, on various arithmetic problems. Attention was paid to their mastery of basic calculation techniques, to invented strategies, and to types of errors displayed in calculation.

Subjects

The American subjects were eight male and eight female second-grade students (mean age = 8.1 years) and eight male and eight female fifth-grade students (mean age = 11.2 years) selected from a public school in Ithaca, New York. Baoulé and Dioula subjects were selected randomly from students in two different schools in the city of Bouaké. There were 16 third-grade Dioula (mean age = 9.8 years); and 16 third-grade Baoulé (mean age = 9.4 years); 16 sixth-grade Dioula (mean age = 12.6 years); and 16 sixth-grade Baoulé (mean age = 12.8 years). Half of the subjects at each age level were boys and half were girls. We selected African students one grade in advance of their American peers in order to equalize number of years in school, as the African schools did not provide a year of kindergarten.

Tasks

Each subject, interviewed individually, was given a large number of arithmetic tasks. The American children were tested in English, and the African children in French, the language in which they received instruction. Here we focus only on written problems. Each child was asked to write down and solve nine addition problems, three of each of the following three types: simple addition where no carrying was required and where no problems of alignment were present (e.g., 62 + 11); problems where carrying was required but no problems of alignment were present (e.g., 57 + 25); and problems where no carrying was required but where potential problems of alignment were present (e.g., 234 + 43).

Results

The results concerning accuracy (Table 14.3) show that older children were highly accurate, achieving almost perfect success in all cultural groups. The results also show that the younger Americans scored somewhat higher than the

TABLE 14.3
Mean Scores by Cultural Group and Grade on Written Addition Tasks

		Overall Written Problems (Maximum 18)	*No Carry–No Alignment (Maximum 6)*	*Carry (Maximum 6)*	*Alignment (Maximum 6)*
American	2nd grade	12.25	5.31	3.13	3.94
Baoulé	3rd grade	9.75	4.81	2.19	2.75
Dioula	3rd grade	9.68	4.31	1.81	3.56
American	5th grade	17.06	5.87	5.43	5.75
Baoulé	6th grade	16.81	5.69	5.81	5.31
Dioula	6th grade	17.62	6.00	5.75	5.87

Baoulé and Dioula. Examination of the means for the younger children shows that all three cultural groups exhibit the same pattern of scores: The simple addition problems are the easiest, followed by the alignment problems, with the carrying problems being the most difficult. Thus, all three cultural groups experience the same sorts of difficulties with the various problems.

Virtually every older child employed the standard written algorithm correctly to solve the problems. A few children made minor execution errors in doing this. Virtually no informal or invented strategies were employed. In the case of the younger children of all three groups the situation was somewhat different. Most frequently (over 50% of the cases) children use standard written algorithms to solve the written problems. On a significant proportion of the trials (around 25%) subjects use a written strategy that is seriously defective. On a small proportion of the trials (about 10%) they give a response labeled as *other,* which generally involves guessing or an ambiguous response. On a similar small proportion of the trials (about 12%), young children, especially Dioula and American, employ informal or invented strategies of one kind or another. Although these strategies are not numerous, they do occur, taking similar forms in both Africans and Americans.

Inasmuch as the older children made virtually no errors, the following analysis concentrates on the errors of the younger children. The errors were coded into the following categories, which were originally derived from observations of American children: (1) *No strategy:* e.g., no answer given; (2) *Counting error:* an error committed in the execution of a counting procedure; (3) *Wrong number fact:* a wrong, memorized answer was used; (4) *Incorrect written numbers:* addends are recorded incorrectly, as in writing *42010* for "*quatre-vingt dix*"; (5) *Incorrect alignment:* written addends are not aligned according to place value, as in writing the 6 under the 2 in 21 + 6; (6) *Defective carrying:* when carrying in written addition is necessary, a defective procedure is used, as in writing both digits at the bottom without carrying 1 to the next column; (7) *Other errors:* any other miscellaneous errors.

TABLE 14.4
Frequency and Proportion of Error Categories in the Three Younger Groups on the Three Types of Problems

		Categories of Error							
Problem Type	*Group*	*No Strategies*	*Counting Error*	*Wrong number Fact*	*Incorrect Written Number*	*Incorrect Alignment*	*Defective Carrying*	*Other*	*Total Number of Errors*
No carry-no align-ment	American	2(22)	0(0)	2(22)	3(33)	0(0)	0(0)	2(22)	9
	Baoulé	5(36)	0(0)	0(0)	7(50)	0(0)	0(0)	2(14)	14
	Dioula	3(17)	2(11)	0(0)	4(22)	2(11)	0(0)	7(39)	18
Carry	American	2(7)	0(0)	2(7)	3(10)	0(0)	10(34)	12(41)	29
	Baoulé	5(9)	3(6)	7(13)	14(26)	2(4)	20(38)	2(4)	53
	Dioula	5(9)	4(7)	7(13)	9(16)	3(5)	20(36)	7(13)	55
Alignment	American	1(7)	0(0)	1(7)	0(0)	10(71)	1(17)	1(7)	14
	Baoulé	2(5)	2(5)	2(5)	11(27)	19(46)	0(0)	5(12)	41
	Dioula	2(8)	1(4)	3(12)	5(20)	12(48)	0(0)	2(8)	25
Overall	American	5(10)	0(0)	5(10)	6(11)	10(19)	11(21)	15(29)	52
	Baoulé	12(11)	5(5)	9(8)	32(30)	21(19)	20(19)	9(8)	108
	Dioula	10(10)	7(7)	10(10)	18(19)	17(17)	20(21)	16(16)	98

Note. The numbers in parentheses indicate the percentage of all errors made by that cultural group for that problem type in that category of error.

The total numbers of errors made by each cultural group reflects the accuracy scores. The young Americans made only 52 errors, as compared to 98 for the young Dioula and 108 by the young Baoulé. (See Table 14.4). However, the *proportion* of each type of error is remarkably similar when the cultural groups are compared. Perhaps the only major difference in proportional error rates occurs in the incorrect written numbers category, which comprises 30% of the total young Baoulé errors and 19% of the total young Dioula errors, but only 11% of the young American errors.

Discussion

The first hypothesis stated that because of language difficulty and unfamiliarity with schooling, young Baoulé and Dioula children perform at a somewhat lower level than Americans on arithmetic tasks. The results support the hypothesis. After a few years of schooling, American children and Africans from both groups performed at roughly the same levels on virtually all tasks. It seems reasonable to attribute African children's initial difficulty to some combination of language problems and lack of familiarity with the institution of Western-type schooling, although we have no direct measures of these variables.

The second hypothesis stated that because of similarities both in general cognitive abilities and in informal mathematical knowledge, Baoulé, Dioula, and Americans should acquire through schooling roughly the same mathematical concepts and procedures. This hypothesis was supported by several features of the data. In the case of written addition, and other tasks too, the clear and consistent finding was that older African children achieved the same high level of success as did older American children. Moreover, the younger African children, who generally performed at a lower level than their American counterparts, nevertheless achieved reasonable levels of competence on several tasks. Also, the data show that African and American children employ similar informal strategies and make similar types of errors.

As maintained above, several factors may explain the overall similarity in mathematics knowledge among the three groups. Americans and Africans seem to share similar cognitive processes. One of these is learning ability or plasticity: Children are adaptable; they learn to adjust to the immediate environment. Even though Western schooling is a cultural anomaly for traditional Africans, they seem to adapt to it quickly and well, even when a foreign language is employed. Hence, the similarity in academic knowledge is partly the result of efficient school learning. Another shared cognitive process, overextension of rules, seems to produce common errors.

Another factor contributing to similar mathematical performance among the groups is common informal knowledge. Children begin schooling with informal notions of arithmetic and use them to assimilate what is taught. Children assimilate the traditional content of elementary school arithmetic into familiar notions such as inequality, numerosity, equivalence, and addition. The result is academic knowledge that is the joint product of assimilation and instruction. In the present study, the three groups' similar performance may in part be attributed to a common base of informal knowledge into which school arithmetic was assimilated.

Because previous research has shown the Dioula to be more adept than the Baoulé at calculation activities, the third hypothesis predicted superiority of the Dioula over the Baoulé with respect to written arithmetic. The results provide moderate support for the hypothesis: In many cases, the Dioula and Baoulé did not differ; when differences existed, the Dioula generally performed at a higher level as predicted.

In general, the results show how easily both Baoulé and Dioula children adapt to Western schooling. These African children acquire mathematical knowledge much as American children do, at least at the levels we have studied. We attribute these accomplishments to plasticity, common cognitive processes, and informal mathematical knowledge shared by the African and American children. The generality of the results can be ascertained only by examining a variety of cultures with different types of informal knowledge and educational training.

Study III: Social Class and Race in America

For many years there has been controversy concerning the cognitive abilities of poor children, and particularly inner-city blacks. All parties to the controversy accept the fact that poor children as a group, and inner-city blacks in particular, exhibit low levels of scholastic achievement (Coleman, 1966). Some writers propose that school failure is a consequence of deficient intellectual skills, which in turn are caused by either an inadequate environment (Hunt, 1969), or an inferior genetic endowment (Jensen, 1969). Other writers (e.g., Cole & Bruner, 1971; Ginsburg, 1972) maintain that poor children's cognitive skills are not deficient but instead are, to some extent, different, and that poor children's school failure must be explained on grounds other than cognitive deficit, for example, black children's perception of limited economic opportunities (Ogbu, 1978).

Despite the heated nature of the controversy, evidence concerning poor children's intellectual abilities is sparse. One exception is Labov's (1972) pioneering research, showing that poor black children's language is not deficient; indeed, it is a complex and rich system, in some respects different from but not inferior to standard English. Unfortunately, work of this quality and importance is not common. In particular, there is hardly any research on poor children's cognitive skills directly related to academic achievement (e.g., mathematical abilities).

In view of the scarcity of data, and the perspectives afforded by cross-cultural research, a study by Ginsburg and Russell (in press) investigated social class and racial differences in early mathematical thinking. The primary aim was to discover whether lower-class children, including blacks from the inner city, are deficient in concepts and strategies necessary for later success in school mathematics. Cognitive deficit theory, whether of the environmentalist or nativist variety, predicts that lower-class children are deficient in basic cognitive processes. By contrast, the work of Labov and our own cross-cultural research suggest that basic cognitive skills are extremely widespread, if not universal, and should be no less prevalent in lower-class Americans than in unschooled Africans.

Design

This investigation, conducted in the Washington D.C. and Baltimore areas, involved children at the prekindergarten and kindergarten levels: black and white, lower and middle class, from intact and single-parent homes. Children were selected according to a 2 (age) × 2 (race) × 2 (social class) design. There were 18 subjects in each of the eight cells, for a total sample of 144 children. Family status was allowed to vary freely within the constraints of this design.

Subjects

Half of the subjects were at the prekindergarten level (average age 4 years, 9 months) and half were at the kindergarten level (average age 5 years, 9 months). All the children attended all-day child-care programs, with some of the kindergarten children attending public school kindergartens for part of the day. (Those not attending public school kindergarten received similar training in their day-care centers.) Half of the children were lower class and half middle class. The social class background of the children was determined by using the Hollingshead two-factor index of social position (Hollingshead, 1957), which assesses social status through educational attainment and occupation of the parent. Levels IV and V were designated as lower class and all other levels, middle class.

Method

Three testers, two female and one male, administered tasks to the children. Before any children at a day-care center were tested, the testers spent a morning or two in the classroom becoming acquainted with the children by taking part in their daily activities. Only after establishing rapport with the children did the testers begin to test individuals. The children were tested in three separate sessions. Among the 17 tasks given were the Perception of More and Addition Calculation, on which we focus here.

The Perception of More task began with six practice trials. The task requires the child to identify, very quickly, which of two sets on a card is larger. The intent is to have the child decide without counting. On the practice trials, the child was given feedback on the correctness of response. This procedure seemed to clarify the task for the vast majority of children. After the practice trials, children were given six trials for each of three types of array (random, length misleading, regular). No feedback was provided. In each type of array, the same numerical comparisons were involved, namely, 5 versus 6; 5 versus 7; 6 versus 7; 6 versus 8; 7 versus 8; and 7 versus 9. Children could thus receive a score of from 0 to 18 correct.

In the Addition Calculation task, the child was first given two pretest problems involving small sums (2 + 1) and (2 + 2). These problems, like the subsequent test problems, were embedded within a story that was designed to unite the two addends through some sort of action. For example, "Teddy Bear has two pennies. He's walking to the store and finds one more penny. How many pennies does he have altogether?"

For Objects Present problems, as the interviewer told a story problem similar to those used on the pretest, she placed objects representing the addends in front of the child in two separate piles, one for each addend. The specific Objects Present problems were (4 + 2), (5 + 4), and (6 + 2).

In Objects Absent problems, the experimenter explained: "Now I'm going to put some things in Teddy Bear's cup. You can use these blocks or your fingers to help you count or you can count in your head to find the answers." As the experimenter mentioned the addends during the story problems, she would quickly drop sets of objects into the puppet's cup in such a way that the child could not count the number of elements in each set nor count the sum, as it was hidden in the cup. Thus, in the Objects Absent problems, the child saw that two sets of objects were combined in a cup, but could not solve the problem by directly counting those objects. Instead, the child had to form a representation of the spoken numbers, with blocks or fingers or mentally, in order to solve the problem. The child was reminded to use fingers or the blocks on each problem if he or she failed to solve it. The Objects Absent problems were (3 + 2), (7 + 3), and (4 + 3). Objects Present problems were always given before Objects Absent problems. Both number of correct responses and the child's strategies were scored.

Results

In the case of Perception of More, a 2 (age) × 2 (social class) × 2 (race) × 2 (family status) × 3 (type of problem) ANOVA failed to show a significant main effect for race, social class, or family status. The means for black and white children were respectively 12.63 and 12.78, and for lower-class and middle-class children, 12.80 and 12.62. The analysis did reveal a main effect of age and of type of problem. The preschool children's means for the regular, random, and length misleading problems were 5.17, 4.80, and 2.21 respectively, whereas the comparable means for the kindergarten children were 5.69, 5.15, and 2.50. These results show that although kindergarten children achieve slightly superior overall performance, both age groups exhibit similar patterns of performance: Children were highly accurate on the regular and random problems, which can be solved by using stimulus information concerning length or area covered, whereas children performed poorly on the misleading problems, where use of length information leads to an incorrect judgment. This pattern of response, observed previously in Africans by Posner (in press), was observed in black and white children, and in middle-class and lower-class children. In brief, the Perception of More results showed no significant social class or racial differences, and subjects used the anticipated strategies, which produced a high degree of success under favorable stimulus conditions.

In Addition Calculation, the ANOVA revealed no significant main effects for race or class. By contrast, there was a main effect for age, with the kindergarten children, who achieved an overall mean of 3.91 (of 6.00 possible), performing more accurately than the preschool children, whose mean was 2.54. A main effect of problem was also significant. As expected, the Objects Absent problems ($\bar{x}$ = 1.11) were harder than the Objects Present problems ($\bar{x}$ = 2.12). Both preschool and kindergarten children were moderately successful on the Objects Present problems, with means of 2.34 and 1.91 respectively. But the preschool children achieved little success on the Objects Absent problems, with a mean of .61, whereas the kindergarten children, with a mean of 1.61, achieved some success.

On the Objects Present problems, the addition strategies used by preschool and kindergarten children were confined almost exclusively to basic enumeration. On the Objects Absent problems, preschool children generally failed to use any sensible strategy whereas kindergarten children often tried to solve the problems by counting representations of the absent objects.

In brief, there are no social class or racial differences in addition calculation. There is normal age-related development, predictable use of counting strategies, and the expected difficulties with mental representation. Other portions of the Ginsburg and Russell (in press) study showed virtually no racial differences and only a few cases of social class differences.

Discussion

This study shows that in the vast majority of cases, children of different social classes and races demonstrated similar basic competence on the various tasks and used similar strategies for solving them. If these competencies and strategies were not evident at the preschool level, they emerged by kindergarten age. These results suggest that although culture clearly influences certain aspects of cognition (e.g., linguistic style), other cognitive systems seem to develop in a uniform and robust fashion, despite variation in environment or culture. Children in different social classes, both black and white, develop similar cognitive abilities, including basic aspects of mathematical thought. The causes of this uniform development are not fully understood. Perhaps poor children's environment is not in fact deprived with respect to mathematical objects, events, and experiences, just as it is not deprived with respect to linguistic objects, events, and experiences. Or perhaps, like language, mathematical thinking is in some sense an innate, maturational phenomenon, requiring only minimal environmental stimulation to unfold naturally. Although this study was not designed to resolve such issues, it does show that the cultural deprivation view is incorrect in its basic empirical assumption, which is that lower-class children show inferior cognitive function. At least in the case of early mathematical thought, it is lower-class competence that needs explaining, not lower-class deficiency.

This study showed almost no racial differences. Even if one were foolishly to grant the existence of race as a pure variable, one finds that it has only trivial effects on early cognitive function, at least in the case of early mathematical thought. Jensen's (1969) notion that lower-class and black children exhibit general weaknesses in abstract thinking is wrong. Many of the tasks used in this study are prime examples of abstract thought yet show no association with racial or social class differences.

Conclusion and Implications

The general conclusion emerging from these three studies is that early mathematical thought displays considerable uniformity despite variations in culture, class, and race. Unschooled Africans seem to organize school arithmetic in the same fashion as schooled Americans. Middle-class and lower-class Americans, black and white, possess various informal mathematical skills. Although the groups studied—Africans and Americans, blacks and whites, middle-class and lower-class—differ in certain ways, the dominant impression is one of common cognitive processes and robust development of systems of mathematical thought.

The research suggests several implications for education. One is that educators must take seriously the notion that upon entrance to school virtually all children possess many intellectual strengths on which education can build. This is likely to be as true of lower-class and of black children as of anyone else. In a sense, without the benefits of schooling, young children already understand basic notions of mathematics. Children know what *more* is, and they possess both concepts and techniques of addition. There is of course much for them to learn, but their biology and culture have given them a good mathematical head start. Elementary education should therefore be organized in such a way as to build upon children's already existing cognitive strengths.

A second implication is that if we fail to educate poor children, then it does not help to blame the victim by proposing that poor children are cognitively deficient or genetically inferior. Such propositions are empirically incorrect, at least insofar as early mathematical thinking is concerned. We need to consider other possible contributions to poor children's observed low levels of academic achievement. These may include motivational factors linked to expectations of limited economic opportunities, inadequate educational practices, and bias on the part of teachers. If factors like these indeed contribute to poor children's school failure, then reform efforts must not be limited to the psychological remediation of the poor child. They must also focus on teaching practices, teachers, and the economic system.

ACKNOWLEDGMENTS

This research was supported by grants from the National Science Foundation (BNS 75-80026), and from the National Institute of Education (NIE-G78-0163).

REFERENCES

Binet, A. The perception of lengths and numbers. In R. H. Pollack & M. W. Brenner (Eds.), *The experimental psychology of Alfred Binet*. New York: Springer, 1969.

Cole, M., & Bruner, J. S. Cultural differences and inferences about psychological processes. *American Psychologist*, 1971, *26*, 866–876.

Cole, M., & Scribner, S. *Culture and thought*. New York: John Wiley, 1974.

Coleman, J. S. *Equality of educational opportunity*. Washington, D.C.: U.S. Government Printing Office, 1966.

Dasen, P. R. (Ed.). *Piagetian psychology: Cross-cultural contributions*. New York: Gardner Press, Inc., 1977.

Ginsburg, H. P. *The myth of the deprived child: Poor children's intellect and education*. Englewood Cliffs, N.J.: Prentice-Hall, 1972.

Ginsburg, H. P. *Children's arithmetic: The learning process*. New York: D. Van Nostrand, 1977.

Ginsburg, H. P., Posner, J. K., & Russell, R. L. The development of knowledge concerning written arithmetic: A cross-cultural study. *International Journal of Psychology*, 1981a, *16*, 13–34.

Ginsburg, H. P., Posner, J. K., & Russell, R. L. The development of mental addition as a function of schooling and culture. *Journal of Cross-cultural Psychology*, 1981b, *12*, 163–178.

Ginsburg, H. P., & Russell, R. L. Social class and racial influences on early mathematical thinking. *Monographs of the Society for Research in Child Development*, in press.

Groen, G., & Resnick, L. B. Can preschool children invent addition algorithms? *Journal of Educational Psychology*, 1977, *79*, 645–652.

Guerry, V. *Life with the Baoulé*. Washington, D.C.: Three Continents Press, 1975.

Hebbeler, K. Young children's addition. *Journal of Children's Mathematical Behavior*, 1977, *1*, 108–121.

Hollingshead, A. *Two factor index of social position*. Unpublished manuscript, 1957.

Houlihan, D. M., & Ginsburg, H. The addition methods of first- and second-grade children. *Journal for Research in Mathematics Education*, 1981, *12*, 95–106.

Hunt, J. M. *The challenge of incompetence and poverty*. Urbana, Ill.: University of Illinois Press, 1969.

Jensen, A. R. How much can we boost I.Q. and scholastic achievement? *Harvard Educational Review*, 1969, *39*, 1–123.

Kiminyo, D. M. A cross-cultural study of the development of conservation of mass, weight, and volume among Kamba children. In P. R. Dasen (Ed.), *Piagetian psychology: Cross-cultural contributions*. New York: Gardner Press, Inc., 1977.

Labov, W. *Language in the inner city*. Philadelphia: University of Pennsylvania Press, 1972.

Lave, J. Tailor-made experiences in evaluating the intellectual consequences of apprenticeship training. *Quarterly Newsletter of Institute for Comparative Human Development,* 1977, *1,* 1–3.

Ogbu, J. U. *Minority education and caste*. New York: Academic Press, 1978.

Posner, J. K. The development of mathematical knowledge in two West African societies. *Child Development,* 1981, in press.

Scribner, S., & Cole, M. Cognitive consequences of formal and informal education. *Science,* 1973, *182,* 553–559.

Scribner, S., & Cole, M. Literacy without schooling: Testing for intellectual effects. *Harvard Educational Review,* 1978, *48,* 448–461.

Sharp, D., Cole, M., & Lave, C. Education and cognitive development: The evidence from experimental research. *Monographs of the Society for Research in Child Development,* 1979, *44,* Serial No. 178.

Stevenson, H. W., Parker, T., Wilkinson, A., Bonnevaux, B., & Gonzalez, M. Schooling, environment, and cognitive development: A cross-cultural study. *Monographs of the Society for Research in Child Development,* 1978, *43,* Serial No. 175.

Zaslavsky, C. *Africa counts*. Boston: Prindle, Weber, and Schmidt, 1973.

15 Learning to Add and Subtract: A Japanese Perspective

Giyoo Hatano
Dokkyo University, Saitama, Japan

In the first sections of this chapter I review Japanese studies on the process of acquiring early skills of addition and subtraction. I divide this process into several steps, describe some of the difficulties children encounter, and suggest teaching methods to overcome them in each step. In the next sections of the chapter I examine linguistic and cultural factors, including the use of abacus, probably contributing to the development of Japanese children's computational competence. Then, in the final section, some suggestions for research on initial learning of addition and subtraction are presented.

DEVELOPMENT OF NUMBER CONCEPT AND CALCULATION SKILLS UP TO KINDERGARTEN

Japanese kindergartens usually include 4-year olds and sometimes 3-year olds, and are not regarded as part of elementary school. I will use the term Grade K to refer to the year just preceding Grade 1. At this age level, about 23% of children go to day-care centers, and 14% stay home all day. Usually no systematic teaching of addition and subtraction skills, nor of the conception of number, is given before Grade 1. A survey by the National League of Institutes for Educational Research (1971) found most kindergarten and day-care center teachers thought "learning of socially approved behavior patterns" and "extending and deepening of experience" were the most important objectives of early education. Almost none replied that the teaching of letters and numbers was important. Of course these were opinions stated publicly, and in fact the teachers might be teaching some number skills because of parental pressure.

Parents want their children to acquire some number skills before Grade 1. According to Fujinaga, Saiga, and Hosoya (1963), more than 80% of the middle-class mothers queried wanted their kindergarten children to master addition, and 70% of them, subtraction of numbers up to 10 before entering grade 1. The corresponding figures for numbers beyond 10 were about 40% and 30%, respectively. The mothers added that they were teaching number skills whenever they found appropriate opportunities. More than 80% of mothers of Grade K children replied that they were teaching addition and subtraction up to 10. Therefore, many young children acquire preliminary calculation skills before entering Grade 1. However, parental teaching tends not to be systematic, but biased toward counting and calculation. As a result, Grade K children's number skills tend to be fragmentary and unbalanced.

Addition and Subtraction Skills

Fujinaga et al. (1963) conducted a series of surveys on number skills of young children in which they gave some addition and subtraction problems. Their results can be summarized as follows: Most Grade K children could find the correct answer to the numerical addition problems up to 5; more than half could do so up to 10. They showed slightly lower performances on the corresponding subtraction problems; they seldom showed the correct solution when regrouping was involved, especially for subtraction. Mitsuyasu, Imaizumi, Ikeda and Yoshida (1971) also found that Grade K children could solve elementary addition and subtraction problems up to 10.

Yoshimura (1974) studied children's strategies for single-digit addition problems and found wide individual differences as well as problem-to-problem variations. Two typical strategies were incrementing (setting the first addend using fingers and counting up the second by fingers) and a strategy heavily depending on addition of equal numbers, for which they had somehow learned answers. Thus $2 + 3 = 5$ because the answer should be one larger than $2 + 2 = 4$.

Conception of Number

At this stage, children do not have a fully developed conception of number in the Piagetian sense. This is partly because they are not given systematic teaching in kindergarten and parental teaching is oriented only to counting and calculation. Many Grade K children fail to compare two sets numerically even after correct cardinal numbers are assigned by counting. For example, in a survey by Nagai, Inagaki, and Hatano (1966) about half the children answered that one of two sets of five elements, arranged differently, was larger than the other, even though they knew both sets had five elements. Conservation of number is harder than the numerical comparison of two sets either by counting or one-to-one correspondence: Only a third could answer correctly, according to the same survey.

Thus, Mitsuyasu et al. (1971) note, it is also true with Japanese young children that apparent number skills like counting and addition–subtraction are much more advanced than their "real" understanding of number assessed by the Piagetian tasks. The data of Fujinaga et al. (1963) also support this conclusion.

Teaching Numbers to Grade K Children

Systematic teaching of number to young children should facilitate their understanding of number as well as their acquisition of addition–subtraction skills. In other words, we should not teach addition–subtraction number facts as if children were learning paired associates. Nor should we force children to learn rote-counting solution procedures and to practice them mechanically. They should learn what addition and subtraction equations represent and be able to confirm the results of calculation by actual manipulation of objects. In order to make this possible, a semiconcrete representation of numbers might be introduced between purely abstract numbers and real things. According to Ginbayashi (1969), one of the leaders of the Sugaku-kyoiku-kyogikai (Council of Mathematics Education, abbreviated as SKK hereinafter), an organization consisting of thousands of mathematics teachers as well as university professors in mathematics and education, in order to be an effective teaching aid, this semiconcrete representation or schema should be: (1) representative of the structure of numbers; (2) manipulatable by children; and (3) easily imaginable mentally by children. In Japan, TILE cut out of cardboard, systematically introduced by SKK, is the most popular schema, though dice patterns and calculation sticks are also used. Figure 15.1 depicts some of the TILE configurations.

Doi (1974) compared several kinds of schemata, including TILE, dot patterns, and calculation sticks, in terms of their intuitive comprehensibility. He found that numbers represented by TILEs and dice patterns were better grasped than others, and claimed that, considering that the decimal system is easily represented by TILEs but not by dice patterns, the former seem to be the best schema.

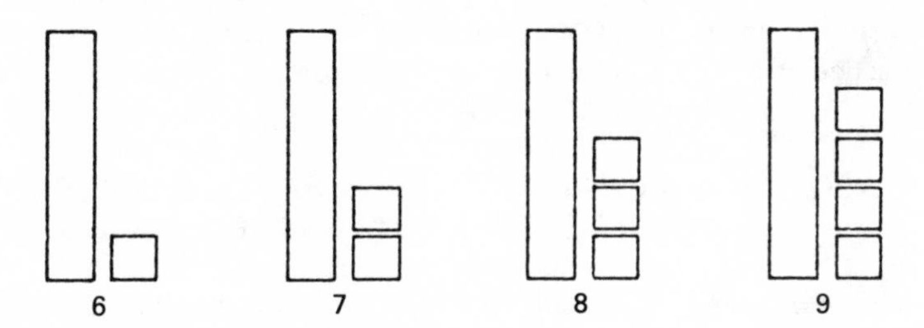

FIG. 15.1. TILEs 6–9 used by SKK.

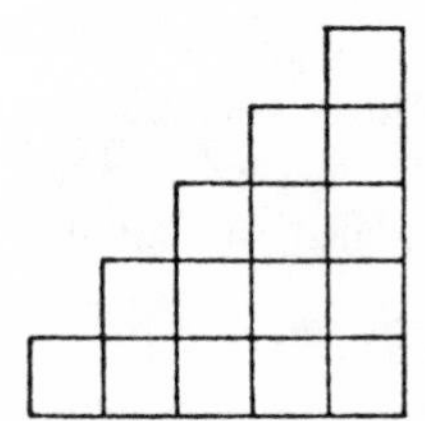

FIG. 15.2. Steplike arrangement of continuous TILEs 1–5.

Before children can understand that a collection of two or more sets assigned the same cardinal number are the same in numerosity, they have to realize that those sets will match the TILE of that number by one-to-one correspondence, and then have to apply transitivity to see that one set will correspond one-to-one with any other set in that collection. For this purpose, SKK proposes to use single TILEs connected by a string, instead of continuous TILEs, as depicted in Figs. 15.1 and 15.2. The next step is to help the children grasp relationships between numbers, so that they can not only judge that k is greater than $(k - 1)$ but also realize that numbers do not change without addition or subtraction. SKK suggests showing children the steplike arrangement of continuous TILEs (Fig. 15.2) and asking them to point out the places where $k + 1$, $k - 1$, or $k \pm 0$, respectively, are satisfied. If children understand this structure of numbers, and "encode" the experimenter's rearranging of the elements of the set as $k \pm 0$, we can expect them to be able to conserve the number.

A series of studies has supported this line of reasoning, especially as embodied in the instructional suggestions by SKK. Only after numbers were associated with the corresponding TILEs or their mental representation could most young children use cardinal numbers reached by counting in numerical comparisons (Ito, 1963; Samejima & Hatano, 1965) and only then could they conserve the number with some direct feedback (Ito, 1963; Ito & Hatano, 1963) or even without any feedback on the conservation task (Samejima & Hatano, 1965). However, practice to strengthen the orientation to respond in terms of number, as well as to encode the actions exerted upon a set as plus, minus, or none, was also necessary to make short-term, intensive training more effective (Samejima & Hatano, 1965; Suga & Hatano, 1968).

Fujinaga, Saiga, and Hosoya (1964) found that dice patterns as schemata were also effective in inducing advanced responses on Piagetian tasks, whereas the sheer practice of counting did not produce any progress.

ACQUISITION OF ADDITION–SUBTRACTION SKILLS DURING ELEMENTARY GRADES

From Grade 1 on, pupils learn math systematically, ordinarily having five school classes each week with each class lasting 40 minutes. Because the Japanese school education system is under the central control of the Ministry of Education, most children learn the same subject matter content at the same grade, irrespective of socioeconomic status, ability, or place of residence. At Grade 1, in the area of "Number and Calculation," objectives include understanding the conception of number and acquiring addition–subtraction skills of single-digit numbers, of tens, and of a two-place number with a single-digit number (Mombusho, 1977). At Grade 2, pupils are expected to learn the decimal system in general and how to add and subtract two- and three-place numbers. Addition and subtraction of positive integers in general, however large they may be, are learned in Grade 3.

In addition to public school, many children have some private lessons, which may influence the development of addition–subtraction skills. Of those areas of private instruction classed as specialized (also including calligraphy, piano, English conversation, etc.), the abacus is one of the most popular. Abacus operation is introduced at Grade 3 at elementary school and is practiced for some hours until Grade 6, but much more extra time is needed for pupils to be able to use the tool effectively. A typical abacus school gives practice of about 1 hour 3 or 4 days a week. An estimated 15–20% of all pupils go to abacus school for some duration, and most of them start learning at Grade 2 or 3, though a few begin 1 or 2 years earlier. Thus, learning the abacus primarily elaborates upon addition–subtraction skills, rather than facilitating their initial acquisition.

A second type of private lesson is given in tutoring schools (*juku*) for general learning, which aim both to prepare able students for entrance exams and to help less competent pupils who otherwise cannot follow what is taught at school. Because Japanese school education is seldom individualized, slow learners often experience difficulty during school hours. These *juku* usually teach all major subjects including math. According to a survey in Kobe, one of the ten biggest cities in Japan, fewer than 10% of first and second graders were going to *juku* in 1973 (Kobe City Institute for Educational Research, 1974), but the proportion is believed to have increased considerably since then.

The following two subsections discuss the stages of acquisition of addition–subtraction ability. In each of these stages children may have some difficulty, and innovative instructional ideas have been proposed to help them overcome it.

Learning Addition with One-Digit Sum and Corresponding Subtraction

SKK claims that ample time should be given to the acquisition of the 100 basic addition and subtraction facts. In fact, it has developed a very careful, systematic program, minimizing the role of counting. The program relies on TILEs in teaching the meaning of addition and subtraction as well as in checking the correctness of the answer. Pupils are allowed to check their calculations by counting the resultant, but the use of a counting-up or counting-down strategy is never recommended. This is because, according to SKK, addition–subtraction by counting often leads pupils to operate on numbers as purely abstract entities, independent of the actual quantities they represent. Therefore, pupils who have been taught according to a counting-oriented program tend to have difficulty in solving concrete or verbally stated problems involving addition–subtraction, though they can solve calculation problems.

Another characteristic of the SKK program is the use of five as an intermediate higher unit. The semidecimal–binary system of number is used instead of the decimal system at the initial stage of learning. Later, this intermediate unit is to be removed, and the unit of fifty is never used. It is interesting to note that the semidecimal–binary system is also used for abacus operation. However, this does not mean SKK has been influenced by this instrument. Rather, SKK teachers found the intermediate unit necessary when they were trying to teach numbers to young children and retarded children. The use of five as an intermediate unit, adopting unsegmented TILEs of five instead of connected or isolated single TILEs, makes it possible to represent the numbers 6, 7, 8, 9 as 5 + 1, 5 + 2, 5 + 3 and 5 + 4 (Fig. 15.1), and thus to grasp them almost intuitively, as different from others.

For learning the so-called "harder" basic facts, (i.e., addition of single-digit numbers with a sum of 10 or larger, and the corresponding subtraction facts) SKK does not recommend counting up or counting down beyond 10. Rather the opinion is that in these calculations the complementary number of 10 should be used with the help of TILEs. For example:

$$7 + 6 \rightarrow (7 + 3) + 3 \rightarrow 10 + 3 \rightarrow = 13.$$
$$14 - 5 \rightarrow (10 - 5) + 4 \rightarrow 5 + 4 \rightarrow = 9.$$

Pupils are not only taught these transformations using TILEs, but encouraged to rely on them when they have any difficulty. For subtraction, double subtraction strategy [i.e., 10 − (5 − 4)] is not prohibited, but neither is it encouraged. Incrementing with subtraction does not occur readily to Japanese.

Yoshimura (1975) found that ten first-graders, half of whom had been tested a year before in kindergarten, tended to use a complementary-number strategy (i.e., breaking down the smaller addend to find the complementary number to 10 (sometimes also to 5) of the larger addend). They no longer employed the counting-up (incrementing) strategy they had previously used. Her conclusion was based on verbal inquiries, but her reaction-time data seem not to support the MIN (x, y) model of Groen and Parkman (1972).

Addition and Subtraction of Multidigit Numbers

In the present teachers' guide (Mombu–sho, 1978), the next step is to learn simpler types of addition–subtraction of two-place numbers, i.e., of tens without ones (e.g., 40 + 20, 70 + 50) and of a two-digit number with a single-digit number (e.g., 43 + 5, 38 + 2). Only after this are children taught the addition and subtraction of typical two-place numbers in which all three terms involved are nonzero two-digit numbers, either with or without regrouping, like 67 − 35 and 41 − 26. SKK strongly objects to this order of teaching. It claims that the next steps should be the typical addition–subtraction of two-place numbers, first without, and then with, regrouping to tens (but not to hundreds).

This claim is based on the SKK metaprinciple, "from the typical to the atypical." In the addition and subtraction of two-place numbers, any problem involving a zero either in tens or ones of the numbers to be processed is regarded as atypical. Therefore, SKK claims, so-called simpler types of problems should be taught later, after children have become familiar with algorithms through typical problems. Otherwise, different solution procedures are used even within addition–subtraction of two-place numbers and many pupils will become confused unnecessarily.

Following the addition–subtraction of two-place numbers, operations upon three-place numbers are taught. After this, addition–subtraction skills can easily be generalized to larger numbers. Again, the conventional order of teaching prescribed in the teachers' guide and that of the SKK program differ. SKK applies the same metaprinciple. According to its supporters, survey findings uniformly showing the difficulty of multiplace subtraction problems that require one or more borrowing operations only demonstrate the deficiency of the conventional sequence of teaching (i.e., "from atypical to typical"). However, the SKK claims for effectiveness of its program are based on "educational practice." Scientifically reliable evaluation studies are lacking.

DOES JAPANESE CULTURE FAVOR THE DEVELOPMENT OF CALCULATION SKILLS?

Speaking generally, Japanese children acquire the skills of addition and subtraction quite well. If we assume that the rate of initial acquisition of these skills is positively correlated with later calculation competence, including multiplication and division, processing of fractions and decimals, a few pieces of evidence have suggested that Japanese children outperform their American counterparts. The International Project for the Evaluation of Educational Achievement (IEA) survey (Husen, 1967) showed that Japanese 13-year-olds performed best out of the 10 countries included on the calculation problems. Umemoto and Haslerud (1959) compared fifth- and ninth-graders living in Manchester, New Hampshire, and in Kyoto, in problem-solving ability as well as preference for arithmetical (calculation and verbally stated arithmetical problems) versus nonarithmetical, verbal (factual knowledge, finding absurdities, etc.) problems. Subjects were presented heterogeneous pairs (one arithmetical and the other verbal) and homogeneous pairs (both arithmetical or both verbal) of problems, and required to solve one of each pair. Japanese children not only chose the arithmetical problems in the heterogeneous pairs more often, but also performed better on both-arithmetical pairs and worse on both-verbal pairs than the American children.

Because school achievement in a subject is primarily a function of time spent on that subject (Berliner & Rosenshine, 1977), this superiority of Japanese children in calculation should be attributed mainly to the larger amount of practice they have either at school or at home. Mild reform in Japanese mathematics education, often inspired by SKK and only partially adopted by the Ministry of Education, may also contribute. However, it should be emphasized that both of the aforementioned factors work within a culture that favors and values calculation.

Preference for Calculation

That the Japanese culture favors calculation can best be substantiated by children's verbally stated preference for mathematics, especially calculation, and by their parents' encouragement of extensive practice in it. Many studies have shown that math is one of the best liked subjects at elementary and secondary schools. According to Shikata (1958), math was liked best or second best both among boys and girls from Grades 1 to 9, except in Grades 7 and 8. A recent study by Hashimoto, Tatsuno, and Fukuzawa (1975), comparing third-graders going to an abacus school with those learning it only at elementary school, showed that for both groups math was ranked second in preference following athletics, and when five areas of math were ranked, the former children chose abacus operation first and calculation second, whereas the latter children chose calculation first and abacus second.

Parental encouragement of children's expertise in calculation is most evident in the case of abacus learning. Abacus school costs about $12 per month, and if the child takes a qualifying examination (most do), additional fees are needed. Moreover, going to an abacus school means practicing calculation at least 100 hours per year. Despite the fact that abacus skills have been losing their utility in Japan, as low-priced hand-held electronic calculators are now widely used in daily life, the number of abacus schools has not decreased. According to the *Yomiuri-shimbun* (May 13, 1971), one of the two best-selling newspapers, more than two million children take some qualifying exam every year. This figure is about one-fifth of the number of elementary school children.

In addition to historical reasons, two explanations can be offered for this preference for calculation. First, calculation skills can be improved monotonously by concentrated practice, and this may be consistent with the traditional Japanese belief in effort. Secondly, learning to calculate may be easy and self-enhancing for Japanese children, because of linguistic factors and the use of the abacus as a tool of calculation. Let me discuss the last two points in turn.

Linguistic Factors

There are at least two linguistic factors in Japanese that may facilitate learning and performing calculation. First, Japanese number words are regular and systematic. This fact may be especially helpful when children learn the addition and subtraction of multiplace numbers (Toyama, 1965) and when they calculate mentally. In Japanese, a number is given orally always from the largest to smallest place, with no irregularities in expressing multiplace numbers. Unlike "eleven," "twelve," and "thirteen" in English, the corresponding Japanese words are, literally translated, "ten–one," "ten–two," "ten–three." "Three hundred and thirteen" is, literally, "three hundred–ten–three" in Japanese (there is no plural form in the language). Therefore, Japanese children tend to have no difficulty in expressing printed numbers orally and no confusion in processing them mentally, once they have understood the place-value system.

Second, the calculation ability of Japanese is much facilitated by *kuku*. Every pupil, at Grade 2, learns to recite short, conventional rhyming phrases each expressing one of the 81 one-digit multiplication number facts very quickly (i.e., $1 \times 1, 1 \times 2, \ldots, 2 \times 1, 2 \times 2, \ldots, 9 \times 9$). These are *kuku,* whose literal meaning is "nine by nine," the last of the multiplication number facts. This certainly helps pupils retrieve the relevant number facts fast and accurately in complex multiplication and division. *Kuku* may have been a product of the cultural emphasis on calculation, but the rhyming phrases are based on linguistic characteristics, i.e., (1) the Japanese language tends to have short syllables because no two consonants come consecutively without a vowel (Umegaki, 1961); and (2) rhyming depends on the fixed number of syllables because the language lacks a stress accent (Bekku, 1977).

Use of the Abacus

An abacus is an external memory as well as calculation device for numbers. Its operation is somewhat complicated, because it has a 5-unit counter and four 1-unit counters for each place value in the decimal system. However, it minimizes the load of mental calculation, for one can "get" the result by moving counters without anticipation. Moreover, when operating on it, one need not internally store the initial nor intermediate results. Therefore, with the instrument, skilled operators can calculate with large numbers quite accurately. But, does learning the abacus improve the ability to calculate by other means?

Hatano, Miyake, and Binks (1977) have shown that very skilled operators could calculate without an abacus as accurately as with the instrument, and often faster. According to the operators' introspection, they were relying upon an interiorized mental abacus. Hishitani and Yamauchi (1976) found that among skilled operators digit span was not correlated with mental calculation ability. They think that those operators were relying upon a mental abacus, or an apparatus visually representing numbers. A study by Hatano and Osawa (1980) showed that with very intensive training, the digit span for this mental apparatus could be extended. Thus both of the two subjects, who had been national champions in the mental calculation division of an abacus tournament, could reproduce rapidly a series of 15 digits either forward or backward. With this mental abacus, they could add or subtract numbers of more than 10 figures mentally.

A study by Ezaki (1980) examined effects of training for interiorization among abacus operators of varying degrees of mastery. Subjects were given 30 easy and 30 difficult mental addition problems with three two-digit numbers. The abacus grade level was a significant predictor of performance. The interiorization training groups also performed better at each grade level than the nontraining groups except at Grade 2 where both groups' performance approached the upper limit. (Grade 2 is the second highest possible grade.) It should be noted, however, that it took students 35 months on the average to reach the level of Grade 2. We can conclude, therefore, that extensive abacus learning improves mental calculation ability, particularly when there is evidence of a mental representation of an abacus.

Abacus learning may improve mental calculation ability even when it is short of developing a mental representation of an abacus, for two reasons. First, the use of an abacus may reduce the anxiety of children facing calculation, thus making the initial learning less troublesome. Written calculation can do the same job, but up to Grade 3, manipulating an abacus is easier than writing digits. The second reason is that extensive practice in abacus operation may help children store and quickly retrieve single-place addition and subtraction number facts. This second conjecture was supported by at least two studies. Asao (1976) found that abacus-learning children made a significantly greater gain than control children on a speeded test of simple computation. A study by Hashimoto et al., (1975) also showed that children going to abacus school got better scores on a test of adding as many consecutive single-place numbers as possible.

The studies just cited relate to extensive abacus learning. In order to assess the possible effects of less extensive abacus learning, Hatano and Suga (in preparation) conducted a study in which they tried to examine the effects of rudimentary abacus learning upon the speed and accuracy of mental and written calculations. Third-grade abacus-learning children were compared with their schoolmates without extra abacus lessons. They were given paper-and-pencil tests of arithmetic, including simple addition and finding complementary numbers to 10 under a strict time limit, three-digit addition and subtraction with carrying and borrowing, verbal problems for which to write equations for calculation, and open sentence problems. The findings were as follows:

1. Abacus learning greatly facilitated the speed of simple computation.
2. It also improved carrying and borrowing.
3. However, it did not show a marked effect on the ability to write and transform equations.

CONCLUDING REMARKS

Based on the preceding reviews of Japanese research on the initial learning of addition and subtraction, I would like to discuss some implications for future investigation. It is clear that Japanese early mathematics education is dependent on an alternative model to counting. Counting-up and counting-down strategies are not recommended in contemporary Japanese mathematics education, especially when regrouping is involved, because these strategies tend to be slower, to be prone to errors, and according to SKK, to lead students to operate on ''abstract'' numbers that are separated from actual quantity. Japanese mathematics education expects students to adopt actual and mental regrouping strategies. This is in contrast to the American emphasis on the progression from the counting-all strategy to the counting-on strategy (see Fuson, this volume).

The Japanese mental regrouping strategy is different both from counting strategies and from the retrieval of stored number facts. The regrouping strategies are based on the extensive use of semiconcrete representations of number. SKK claims that only one kind of representation (i.e., TILE), should be used throughout instruction, instead of mixing several different kinds of representation. In addition, an intermediate higher unit of five is used. SKK believes that by doing this, numbers are equated immediately to the corresponding standard representation of TILEs, which are easily manipulated actually or mentally, so the representation of the resultant number can be subitized.

This conceptual model for addition and subtraction and the corresponding method of teaching early mathematics are closely associated with Japanese culture (e.g., the use of the abacus). Therefore, they may not be applicable to other cultures without considerable modifications. However, their apparent success in Japan will certainly stimulate mathematics educators to systematically examine their applicability.

We can also reasonably conclude that Japanese children's high achievement in mathematics revealed by IEA and other studies is attributable mainly to the Japanese cultural emphasis on computational competence. The time-consuming practice in abacus operation, in the age of electronic computers, seems to be possible only in this cultural context.

Practice on the abacus does not guarantee improvement in the ability to solve verbal and open sentence problems. However, we should not draw the hasty conclusion that the emphasis on computational competence is out-of-date and may inhibit the development of mathematical problem-solving ability. The computational competence achieved through practice on abacus or paper and pencil calculations may convince students, especially those less bright, of their ability to solve a problem analytically or of the utility of analytic problem-solving procedures in general.

Toyama (in Toyama & Ginbayashi, 1971) suggested that, on one hand, practice in computation should not be mechanical but should instead be based on a full understanding of the meaning of the operations, but that on the other hand, computational skill is instrumental in problem-solving ability. According to Toyama, computation ability (the ability to apply accurately an appropriate strategy) is an important component of analytic ability (the ability to choose an appropriate strategy from among the various strategies available and to perform it accurately). It is quite possible that the success of Japanese children in IEA should be attributed to their computational competence, and that their apparent success in verbal problems was due to their having a longer time to solve these problems, because they had spent a shorter time on the computational problems. However, it may also be asserted that, following Toyama, computational competence achieved in early mathematics led Japanese children to develop their analytic problem-solving ability in general. We might wish to conduct a cross-national research on mathematical problem-solving ability, especially in the context of daily life situations.

REFERENCES
[In Japanese]

Asao, H. [Effects of abacus learning on children's intellectual development.] *Nihon-shuzan* (Abacus in Japan), July 1976, No. 269, 18-22.

Bekku, S. [*Rhythm of Japanese language.*] Tokyo: Kodansha, 1977.

Berliner, D., & Rosenshine, B. The acquisition of knowledge in the classroom. In R. C. Anderson, R. J. Spiro, & W. E. Montague (Eds.), *Schooling and the acquisition of knowledge*. Hillsdale, N.J.: Lawrence Erlbaum Associates, 1977.

Doi, S. [A study on the formation of number concept.] *Shinshu University Bulletin of Education,* 1974, *32,* 49-64.

Ezaki, S. [On interiorized activity in abacus-derived mental arithmetic.] *Nihon-Shuzan* (Abacus in Japan), April 1980, No. 314, 2-5.

Fujinaga, T., Saiga, H., & Hosoya, J. [The developmental study of the number concept by the method of experimental education: II.] *Japanese Journal of Educational Psychology,* 1963, *11,* 75-85.

Fujinaga, T., Saiga, H., & Hosoya, J. [The developmental study of the number concept by the method of experimental education: III. The preliminary research.] *Japanese Journal of Educational Psychology,* 1964, *12,* 44-53.

Ginbayashi, H. [Mathematics and math education.] In K. Hatano & H. Ginbayashi (Eds.), [*Logic and psychology of school subjects: 4. Mathematics.*] Tokyo: Meiji-tosho, 1969.

Groen, G. J., & Parkman, J. M. A chronometric analysis of simple addition. *Psychological Review,* 1972, *79,* 329-343.

Hashimoto, J., Tatsuno, C., & Fukuzawa, S. [Effects of abacus learning upon classroom performances.] *Shuzan-shunju* (Abacus Quarterly), November 1975, No. 40, 2-20.

Hatano, G., Miyake, Y., & Binks, M. G. Performance of expert abacus operators. *Cognition,* 1977, *5,* 57-71.

Hatano, G., & Osawa, K. *Digit span of abacus-derived mental arithmetic experts.* Paper presented at the 3rd Noda Conference on Cognitive Science, 1980.

Hatano, G., & Suga, Y. *Abacus learning, digit span and computational skills among third-graders,* in preparation.

Hishitani, S., & Yamauchi, M. [An analysis of information-processing system of skilled abacus operators in mental arithmetic.] *Kyushu University Bulletin of Education (Educational Psychology),* 1976, *20*(2), 55-62.

Husen, T. (Ed.). *International study of achievement in mathematics: A comparison of twelve countries.* New York: John Wiley, 1967.

Ito, Y. [An experimental study of children's conception of number: The conception of equivalence of sets and conservation of number.] *Japanese Journal of Educational Psychology,* 1963, *11,* 157–167.

Ito, Y., & Hatano, G. An experimental education of number conservtion. *Japanese Psychological Research,* 1963, *5,* 161–170.

Kobe City Institute for Educational Research. [A survey on children's study activities at home.] *Research Report,* 1974, no. 137.

Mitsuyasu, F., Imaizumi, N., Ikeda, A., & Yoshida, M. [A study on the development of number concept among kindergarten children: Phenomenal and real concepts of number.] *Fukuoka University of Education Bulletin (Part 4),* 1971, *20,* 113–126.

Mombu-sho (Ministry of Education). [*The elementary school course of study,*] 1977.

Mombu-sho (Ministry of Education). [*Elementary school teachers' guide: Mathematics,*] 1978.

Nagai, T., Inagaki, K., & Hatano, G. [Development of number abilities in 4- and 5-year old children.] *Monograph of Japanese Institute of Child Research,* 1966, *6,* 1–18.

National League of Institutes for Educational Research. [*Opinion survey for the improvement of compulsory education,*] 1971.

Samejima, Y., & Hatano, G. [The acquisition of counting operation as a method of quantification.] *Japanese Journal of Educational Psychology,* 1965, *13,* 234–246.

Shikata, J. [*Psychology of mathematics learning.*] Tokyo: Meiji-tosho, 1958.

Suga, Y., & Hatano, G. [*The role of transformational concept in the acquisition of numerical comparative judgments.*] Paper presented at the 10th Annual Convention of Japanese Association of Educational Psychology, 1968.

Toyama, H. [*How to teach mathematics.*] Tokyo: Meiji-tosho, 1965.

Toyama, H., & Ginbayashi, H. [*Foundations for updating math education: 2. The water supply method.*] Tokyo: Kokudo-sha, 1971.

Umegaki, M. [*Introduction to comparative study of English and Japanese languages.*] Tokyo: Daishu-kan, 1961.

Umemoto, T., & Haslerud, G. M. [U.S.-Japan cross-national study of children's responses and attitudes to arithmetical problem situations.] *University of Kyoto Bulletin of Education,* 1959, *5,* 69–96.

Yoshimura, T. [*Strategies for addition among young children.*] Paper presented at the 16th Annual Convention of Japanese Association of Educational Psychology, 1974.

Yoshimura, T. [*Strategies for addition among young children (2nd report).*] Paper presented at the 17th Annual Convention of Japanese Association of Educational Psychology, 1975.

16 The Psychological Characteristics of the Formation of Elementary Mathematical Operations in Children

V. V. Davydov
The Institute of General and Pedagogical Psychology of the Academy of Pedagogical Sciences of the USSR, Moscow

Children's general orientation to the world of mathematics is largely a function of the initial concepts and operations they master. This orientation, in turn, has an important influence on children's further mathematical development. Identifying the concepts and skills with which the study of mathematics in school should begin is, therefore, of considerable importance in the improvement of mathematical instruction because these concepts and the operations connected with them form the foundation for the structure of the entire academic discipline. In our view, many difficulties in teaching mathematics in schools[1] arise from the fact that neither the content of elementary mathematical concepts nor the methods of introducing them in instruction have yet been subjected to careful psychological investigation.

When teachers choose the initial topics for mathematics instruction, they are often governed by these considerations:

1. A primary purpose in elementary school is to teach knowledge and skills useful in everyday life. Among the skills a child may employ to solve practical problems are the addition and subtraction of numbers.
2. Addition and subtraction of numbers is close to the experience of children at the age of 6 or 7. Thus, they should be able to master these skills quickly and easily at the inception of school instruction.

[1]Soviet schools have three stages: (1) elementary school, which consists of grades 1–3, with children 7–10 years old; (2) middle school, grades 4–9 (10–15 years); and (3) the final stage of school, grades 9–10 (15–17 years).

That is to say, the structure of traditional instructional programs is based on the principle of immediate practical utility. Over a long period of time, practitioners of this traditional approach have become so proficient in teaching the mechanical skills of addition and subtraction to 6- and 7-year-olds that there is now little room for improvement in their programs and methods of instruction.

On the other hand, our experimental work in curriculum development has made us aware that the concepts and skills with which we begin school mathematics instruction must be specially chosen. In particular, we pose the following question: Does the traditional instruction described in the foregoing lead children to an *understanding* of the intrinsic bases of these skills as mathematical phenomena? The observations of many psychologists demonstrate that this is unfortunately not the case. Although children of ages 6 or 7 can readily solve in an abstract (e.g., $3 + 2 = ?$ or $8 - 5 = ?$) or concrete (e.g., add two apples to three apples) form, they cannot explain what numbers are, how they arose, or why in using numbers it is necessary to add or subtract. In other words, the ability to solve practical problems does not necessarily imply children's knowledge or understanding of these deeper principles. However, it is important for the development of children's mathematical thinking that they understand the general prerequisites and conditions of the origins of numbers and arithmetical operations. Only then can children consider numbers and operations from a theoretical perspective, gradually becoming accustomed to the nuances of the processes used to generate mathematical abstractions.

THE ORIGINS OF THE NUMBER CONCEPT

From a logico-psychological point of view, a person's true understanding of a subject can be shown by the ability to reproduce and demonstrate to another person the entire process of its origin. In the case of the concept of number, this means that a student should be able to demonstrate independently to a teacher, using appropriate actions upon objects, why it is both possible and necessary to form this concept. Further, the student should also be able to utilize the numerical properties of *any* quantifiable set for *any* specified purpose. For example, whether or not a child understands the concept of number can be shown by the proper execution of tasks like the following:

1. Require the child to pour into a second container the same amount of water provided in a first container that differs in form from the second. (The first container is a narrow, graduated cylinder, the second a wide-mouthed glass.) (Fig. 16.1). A child who can really isolate the conditions for obtaining a number, that is, who really understands its meaning, should use some intermediate measure, such as a small glass, to determine the amount of water the narrow cylinder

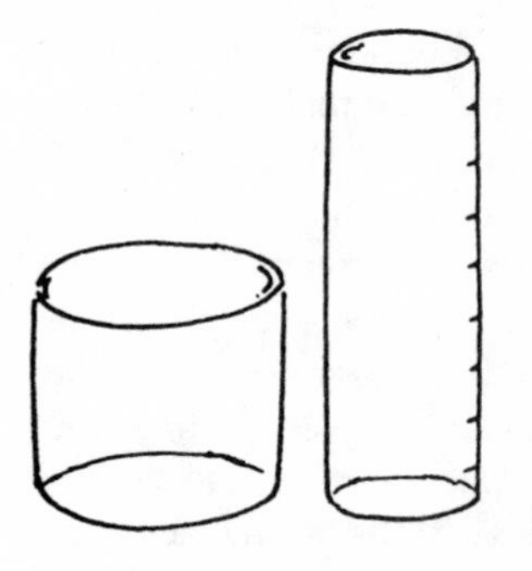

FIG. 16.1.

contains (for example, five small glasses) and then pour the same number of glasses into the wide-mouthed glass.

2. Require a child to determine how many large glasses of water are contained in a series of three large and four small glasses if a small glass is equal to one-half of a large one. (Fig. 16.2). Here, the child must count two small glasses as one large one and obtain the result of five.

3. Using a single set of blocks, require a child to determine various conditions under which several different numerical attributes would be defined. In this task, the child must construct equal groups of blocks and then use those groups as a unit of measure to determine different numbers. (Fig. 16.3). For instance, if 24 blocks are grouped by twos, then the number 12 will be expressed; if grouped by fours, then the number will be six; and so on.

4. Require a child to show how, using a single volume of water in a glass, different numerical descriptions of that same volume of water can be expressed. This task is similar to Task 3 but uses a continuous quantity instead of discrete objects. Different measures (for example, different sized small glasses) must be used to determine several different numbers.

For each of these tasks, the child must recognize the multiple relationship that can exist between a continuous or discrete object (as expressed by its numerical

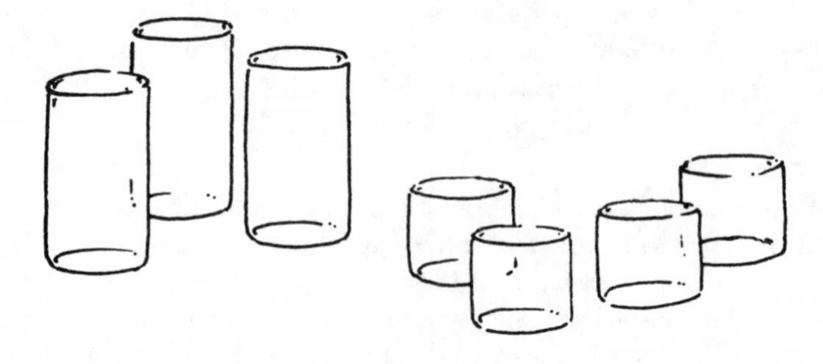

FIG. 16.2.

measure) and some part of that object that has been used as the unit of measure. In so doing, it is of particular importance that the child realize the arbitrary nature of the size of the part (the unit of measure) that is used to determine the measure of the entire object. When measuring, the child should be able to exchange one unit size for another and thereby determine different measures for the same object. In this exercise, the child needs a clear understanding of the *origin* of numerical measure to generate various concrete numerical representations of the object. Only when a child can carry out these fundamental steps can one speak of that child's *understanding* of number as a general mathematical method of expressing quantitative relationships within and between objects.

In our studies we have used a variety of tasks like the preceding to determine the actual understanding of the concept of number in children ages 6 to 8. Later we will discuss the solutions generated by children who received a special instructional program on the meaning of quantity and measurement. Initially, we found that a majority of children enrolled in traditional programs could not carry out these tasks. For instance, in the first and fourth tasks they had no idea how to proceed. In the second task they counted each glass, large or small, as a separate unit and thus obtained an answer of seven rather than the correct response of five. In the third task they counted the blocks singly to obtain 24 and were not able to group out any other unit of counting.

In other tasks these children correctly counted discrete units of objects such as the number of toy cars placed on a table before them. Although these children were able to use a limited concept of number to deal with day-to-day and school problems, they really did not exhibit a true mathematical understanding of the number concept. They did not possess this true understanding because teachers had used "familiar" numbers as the starting point for instruction within the traditional program. On this basis first-grade children quickly progressed to addition and subtraction of numbers known to them only on an experiential bases. However, these numbers do not have for the children the form of a mathematical concept. (We have set forth the logico-psychological theory of concept as it applies to teaching children mathematics and other subjects in several special publications, e.g., see Davydov, 1972.)

First- through third-grade children in the traditional programs exhibited mastery of the standard algorithms for addition and subtraction of single and multidigit numbers (e.g., $8 + 5 = 13$; $26 + 9 = 35$; $134285 - 49$; etc.). However,

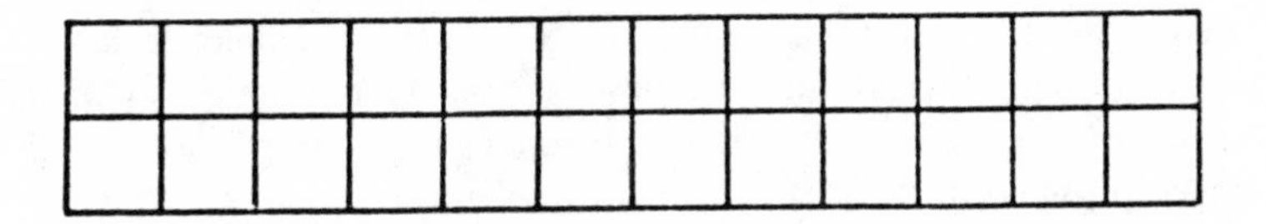

FIG. 16.3.

they were immediately perplexed when asked what possible sense they might make out of the unexpected number sentence "3 + 4 = 5." This number sentence was derived from the second task described earlier. The addends represented two unequal units of measure, two different kinds of physical objects, whereas the sum of 5 corresponds to the correct solution to the task.

Thus the tendency of teachers to provide their children with only a limited exposure to the most familiar, comprehensible mathematical concepts and skills does not guarantee the formation of more general and correct concepts and operations. Although everyday experiences may equip a child with routinized knowledge and actions, they do not explain the origins of these routines. As a consequence, the child does not learn how to describe actively the notions underlying commonplace knowledge and actions, which in the present context are the fundamental mathematical concepts and operations.

If they wish to teach mathematics on the basis of enabling children to demonstrate the sources of the origins of its concepts, then teachers should themselves first determine these sources and begin instruction with them. To do this, however, demands a special logico-psychological analysis of the genesis of the concepts and operations in question. With respect to the addition and subtraction of numbers, an examination of the sources of the concept of number is required. Young children must become acquainted with these sources before numbers and numerical operations are introduced to them.

THE BASIC CONCEPT OF QUANTITY

We have recognized that the basic concept underlying the domain of real numbers studied in middle school is *quantity*. Included within the domain of real numbers are the whole numbers and fractions studied in elementary school. Such important mathematicians and teachers as Lebesque (1936) and Kolmogorov (1960), among others, have described in detail the fundamental role of the concept of quantity in school mathematics programs. They believe the concept of number arises within the context of measurement of a continuous quantity so that a multiple relationship is established between that quantity and a part of it that is used as a unit of measure. According to this it is possible to consider counting as the measurement of a set of discrete objects. Thus, the fundamental question is, "What is quantity?"

Comparing the elements of sets of similar objects and applying the relationships "equal to," "greater than," or "less than" (for example, to lengths of segments or temperature of gases) gives meaning to the notion of quantity. Objects themselves can be considered as quantities when criteria are established to permit application of the trichotomy law to elements A and B of that object. Thus, it is possible to quantify such entities as hardness or the occurrence of events in time.

The Soviet scientist Kagan (1963) writes: "When we determine the criteria of comparison, we change a set into a quantity [p. 101]." In practice, a quantity is usually designated not as a set of elements but as a new concept introduced to help differentiate the criteria for comparison. In this regard, Kagan goes on to say: "In mathematics a quantity is completely determined when a set of elements and the criteria of comparison are indicated [p. 107]." From the standpoint of interest in developing an understanding of the origins of number, it is most important to recognize that the series of natural numbers is an example of a mathematical quantity. Furthermore, when appropriate criteria are established for ordering of rational and irrational numbers, these also may be considered as quantity. Doing so for the irrationals is the foundation of contemporary analysis.

This structure of order we have been characterizing is one of three fundamental structures of mathematics (Bourbaki, 1948). Order, however, is a relationship involving only two elements. Kolmogorov (1971) extended this notion in his formulation of the axiomatics of quantity by including properties of addition and subtraction as well. Although we do not list Kolmogorov's axioms here, we do note that they define completely the modern concept of the system of positive scalar quantities, within which system the choice of some quantity l as the unit of measurement permits the representation of all other quantities by $Q = l \cdot \alpha$ where α is a positive real number. Thus, real numbers are quantities. This representation allows us to think of real numbers as expressing the multiple relationship of quantity, an important point referred to earlier in this chapter.

When working with quantities, one can set up a complex system of transformation through which relationships among properties of quantities can be produced. In producing transformations, one may move easily from equality to inequality as well as perform addition and subtraction with the commutative and associative properties as a guide in addition.

Before turning to the curricular implications of this theory of quantity, one more point merits special attention. The basic properties of addition possess an algebraic structure, expressing in particular the law of composition by which the relation between two elements determines a unique third element as a function. Thus quantity can be characterized both in terms of the order relation and of a function. These are extremely important general mathematical relationships.

CURRICULAR IMPLICATIONS

Given the fundamental importance of the quantity construct, we have concluded that it is desirable to fashion the elementary mathematics curriculum using it as the basis. Mastery of the properties of this concept makes a child aware of the conditions for the genesis of the natural numbers. Furthermore, a later and deeper acquaintance with quantities allows children to enlarge their cognizance of number in a consistent and logical manner to include the whole numbers, rationals, irrationals, and the reals. Although it is true that other tasks face the contemporary school as well, it is necessary from a psychological point of view to show children the bases of the concepts involved.

At any rate, the ultimate aim of instruction in mathematics should be clear from the very beginning. Of course, this does not imply that one must start out to fulfill this aim immediately; rather the intent is to have the aim prescribe the particular form of development of the entire school mathematics program, even in its initial stages. Once children begin to master mathematics, it is important that they understand the need for its fundamentals and the sources of its origins. When adults use mathematical concepts for exclusively practical purposes, they tend to forget the sources of those concepts. This is true of both mathematicians and teachers. When teachers forget the sources of mathematical concepts, a course in mathematics can become defective and fail to help schoolchildren form the fundamentals of true mathematical thought. Kolmogorov (1960) emphasized this very strongly, saying: "Divorcing mathematical concepts from their origins, in teaching, results in a course with a complete absence of principles and with defective logic [p. 10]."

It is not an easy matter in teaching to take into account the origins of concepts. A school subject constructed with this in mind has a special structure distinct from the traditional approach. This is true because a significant portion of the course is specifically designed to lead a child to experience how a concept came about. Acquainting small children with the properties of quantity fulfills exactly the function of introducing them to the process of how the concept of number came about. Only when there is such an introduction can children begin to understand the concept of number and, in particular, can they successfully solve the aforementioned problems.

These considerations guided us in the construction and testing of an experimental mathematics program for grades 1–3. A primary aim was to have children study the fundamental properties of quantity before they became acquainted with numbers. Because lecturing about these properties to 7-year-old children is senseless, we wanted to find a way of working with educational materials so that the children could, on the one hand, see the properties in the things around them and, on the other, learn to record these properties in a definite symbolic system and carry out an elementary mathematical analysis of their relationships. Thus the program contains, first, a listing of those properties of quantity that are to be mastered; second, a description of materials necessary for mastery; and third (and this is the most important from a psychological point of view), a description of actions by which a child may isolate and master the properties of quantities.

The description of the program begins with a brief characterization of the content of the six initial topics that children master in the first half year of the first class (7 to 7½ years of age).

Topics 1 and 2 have as their aim that children, working with manipulatives, determine for themselves which attributes of these objects have the three specific relationships that define quantity. A primary objective is for the child to master the techniques of recognizing these attributes and the symbolic ways to describe their relationships mathematically (by letters, formulas). The children construct a special mathematical "subject" and begin to study its characteristics. In our program this subject takes the form of equalities and inequalities that children discover as they compare and order various physical objects.

Topic 3 studies properties of equalities and inequalities such as transitivity and the reflexivity of equality. This topic acquaints children with particular properties of quantities within a definite written system for representing equality and inequality. Thus children move farther and farther away from direct observations of relationships between concrete objects. They begin to evaluate and express relationships in a logical verbal manner using statements of the type, "if . . . ," and "if . . . , then . . . ."

During Topic 4 children meet the operation of addition (subtraction). They are taught to observe changes in the concrete values of quantities, compare new values with old ones, record the results of this comparison as an "increase" or "decrease," represent the increase or decrease with the symbols "+" and "−," link equalities and inequalities with their properties. and shift from equality to inequality via the operations of addition and subtraction.

Topic 5 deals with the shift from an inequality of the type $A > B$ to equality through addition (or subtraction).[2] This topic leads children to realize that an inequality between quantities can be removed by determining the concrete value by which they differ. This acquaints students with the simplest form of an equation and deepens their understanding of the connection between addition and subtraction.

Finally Topic 6, "The Addition of Equalities and Inequalities," completes the beginning section of the course. It unifies the content of the preceding topics and shows the possibility of representing the value of a quantity as the sum of several values, that is, the possibility of substituting one form of an expression for another. All of this constitutes the prerequisites to familiarizing a child with the commutative and associative properties of addition.

After mastering the content of these six topics, in the second half of the first year children can shift to studying the origins of number as a particular type of quantity. Consequently, the first section of the program has introduced children in a logical, detailed way to problems concerning the origins of elementary mathematical concepts and operations.

Introducing the content of addition and subtraction operations is no trivial matter. A consistent application of these operations first takes place during the problem solving in Topic 6. We note that before the operations of addition and subtraction are mastered, several stages have been covered. First the teacher demonstrates to the children how an object can change with respect to some attribute. This is easiest to show by changing the amount of water in a flask or the weight of a load. The teacher can use various examples from everyday life, all of

[2]Initially Cyrillic letters were used in working with children.

which can be summed up in the statement: "There was so much, and now there is this much." In doing this it is important to show that there are two directions of change—increment and decrement.

The next step consists of describing the change. The children compare the amount of water in two identical flasks and write the expression $A = B$. Then the teacher pours a quantity of water into a new flask and has the children designate this amount by the letter C. The teacher asks the children to write $C > B$. But how can this new amount be obtained? Can C be obtained from A? How can you write down what happens to A? In one form or another the children show that so much water is added to A and that C is the result. The teacher helps the children write down the equation $A + K = C$ and explains the meaning of the symbol = and the letter K. Then the children put the sum in an expression of inequality and obtain $A + K > B$. In a like manner they obtain the expressions $A - K < B$, $A < B + K$, and $A > B - K$. Tasks corresponding to these expressions can be easily executed with attributes of various other materials such as, for example, the weight of sand or salt.

At the next stage, children, using some visual aid, should carry out tasks corresponding to expressions presented by the teacher. Thus for $B = D$ and $B < D$, the children determine the direction of the change and show it with strips of paper. Parallel with this action, the teacher verbally discusses the process: "The sides were equal. If the left side is now the smaller, that means the right side has become more. It has been increased."

After several lessons a new task is given: What must one do in order to restore the equality? We have found from our practice that approximately one-third of the children can immediately and independently give the answer: "It's necessary to take something away" (if something has been added), or "It's necessary to add" (if something has been taken away). The children validate these solutions with the teacher and substantiate the possibility of changing the other side of the inequality. An equation of the type $A + K = B + K$ replaces the former expression of inequality. Moreover, a significant number of the children can independently determine to what extent the measure has changed.

From lesson to lesson, it becomes less and less necessary for the children to use direct aids. Tasks can now be carried out primarily on the basis of verbalizing possible relationships, given certain conditions. At this stage, we often encounter instances of children's reasoning that demonstrate their understanding of the actual connection between various relationships. For example children might say: "A was equal to B. B was decreased. It became smaller than A. That means A has become larger. It is necessary to use the sign 'larger than'."

Once the children are acquainted with the operations of addition and subtraction, we can introduce problems having letters in their conditions. For example, paired expressions of the type $A = B$, $A + K$ ______ B; $A = B$, $A + K$ ______ $B + K$; $C = G$, C ______ $G - D$; etc. Our children correctly solved these types of problems by analyzing literal expressions without recourse to manipulative materials.

The work of children in Topics 5 and 6, as they master the transformation of inequality to equality, has great importance for clarifying the meaning of the operations of addition and subtraction. This work was done in the following manner. Having obtained the result of a comparison and represented it by the expression $A < B$, the children performed a new task. They transformed this inequality into an equality. Using a visual aid, many children independently showed how this could be done. They suggested that B be decreased or that A be increased, and demonstrated this with small wooden or paper strips. Then the teacher showed how such a transformation could be written down: $A < B$, $A = B - \square$, $A + \square = B$.

The boxes in these equations represent that which is added or taken away in order to obtain an equality. But what exactly is added or taken away? While showing a flask of water or the weight of a certain object the teacher explains that although we do not know ahead of time the magnitude of the difference between A and B, we still can express it with the help of x. This symbol stands for the unknown quantity that can be used to transform an inequality into an equality. If $A < B$ then $A = B - x$ and $A + x = B$.

Children, as a rule, quickly understand the meaning of this symbol. Thus in the very first lesson, many of them correctly explain that one cannot add or subtract just any weight or volume. They point out the necessity of knowing the difference between A and B. Several succeeding lessons were given over to working with visual aids in order to acquaint the children with methods of determining this difference. In doing this, it was important not only to show the actual method of action (superimposing strips, pouring water into flasks, etc.), but parallel with this it was important to teach the children to describe with symbols the process and results of the operations with objects.

This is perhaps the most difficult point of the entire topic. This is so because the results of $x = B - A$ present a child with a new meaning for subtraction. Here it means not an actual decrease, as was the case in the preceding topic, but the formal description of the process of comparing the magnitude of B and A. In other words, B remains the same in the physical sense as it was before. The amount that correspond to x has to be obtained from some other physical source. The equation is only a formal way of describing the process of obtaining x. After determining x (the difference), the children "added" it to A and obtained the required equality. The majority of children immediately understand the meaning of the expression of this addition in letters.

All the work in setting up and solving equations using addition and subtraction is carried out with visual aids and expressed in a special system of literal expressions:

$A < B$	(the initial condition)
$A + x = B$	(the planned transformation)
$x = B - A$	(determining the difference)
$A + (B - A) = B$	(the actual equalization)

In further work, the role of physical transformations is gradually reduced and the transformation of the expressions themselves takes on particular significance in the solution of equations. To prepare children for this shift an intermediate strategy of graphic representation is used. The children represent the physical quantities with two line segments A and B. How the difference in measures of the quantities can be determined is then discussed. Line segment A is superimposed on line segment B. The difference, expressed in the form $B - A$, is defined as being equal to x. Then the children, with the help of the teacher, actually measure the quantities and determine the measure equal to x. Then this measure (actually the straight line segment) is added to A and the resulting expression is written down.

Work with an intermediate expression can precede such a shift. This expression fulfills a distinctive function: Having written down an equation, children more often than not seem to forget the initial inequality. Therefore, it is necessary to return to the initial expression, but this time starting from the equation: if $B + x = D$, then $B < D$. With this review of the expression one can, as it were, connect the equation with the inequality in order to move to the subtraction of the smaller quantity from the larger. Thus, all the work takes on the following form:

$A > B$
$A = B + x$
$A > B$
which means
$x = A - B$

Gradually children begin to determine x without the help of any concrete objects or their graphic analogs, using instead a theoretical examination of the relationship of the parts of an inequality. The values that are found are then substituted in an equation. Parentheses help children to do this. They allow children to understand a difference as some actual value.

Work in Topic 6 synthesizes much of the earlier knowledge about the properties of relationships. When the teacher gave problems requiring validation of the results of addition or subtraction of equalities or inequalities, he or she did not, of course, present to the children the formal rules of a systematic algebra course. Rather, the important aim was to foster in the children the ability to use the simplest argument based on the properties of relationships, the ability to arrive at elementary expressions from the point of view of their meaning and not from the external combination of specific conditions.

Thus working on this topic, children solve problems of the following type.

$A = B$	$A > C$	$E < B$
$K > M$	$N = D$	$M < G$
$A - K$ ______ $B - M$	$B \pm N$ ______ $C \pm D$	$E + M$ ______ $B + G$

Work on the substitution of two, three, or more addends for the value of a quantity occupies a definite place in the sixth topic. In a series of special exercises children "expand" or "contract" literal expressions on the basis of given conditions (for example, they rewrite the inequality $A > B$, given the fact that $A = K + M + N$).

All of this provides children with a good preparation for the commutative and associative properties of addition. In the process of working with the simplest literal expressions, children display a lively inclination towards reasoning, mental comparisons, and logical assessments of specific dependencies leading to the direct execution of arithmetical operations.

EVALUATION

Because we assume that the content of the program we have constructed is important for the child's further development in mathematics and especially for the mastery of the meaning of addition and subtraction, an important question arises: How long does this knowledge remain in the child's consciousness?

To answer this question, special tests were administered at the end of the first grade and at the beginning of the second grade. Monitoring the mastery of the content of our program and the forms of mental action developed was an essential part of our investigation. Tests were constructed to measure not only how well children can use the material, but also the degree to which they understand it. In these tests children encounter for the first time problems where familiar mathematical relationships and properties of operations are expressed in an unusual form. The solution of these problems requires that children actually understand the material and correctly interpret situations that reflect the consequences of relationships and properties with which they are already familiar. In particular cases the pupils receive especially difficult problems. The peculiarities of their solutions allow us to judge their understanding of those relationships of quantities that form the basic content of our course.

The majority of the pupils solved the more routine problems correctly, indicating adequate mastery of the content of the introductory sectisn of the program. The results of a series of different problems given to the children at the very beginning of the second grade were especially interesting. These problems differ outwardly from those the children had in the first grade, where they did not work with such "complicated" formulas. Several required that several properties of quantities be taken into account. Examples of these problems are presented in Table 16.1.

Problems 1 and 2, which required a shift to equality by two methods, were solved successfully by the majority of pupils. Problem 3 was also correctly solved by almost all the children (here it was necessary to increase the right member or decrease the left member of the equality). The worst results were

TABLE 16.1
Sample of Unfamiliar Test Problems Presented to Second-Graders

$B - C > K + M$	$A - B + C = D - M$
1. _____ = _____	3. _____ < _____
2. _____ = _____	4. _____ = _____
$A + M = B$	$5 + 2 = 7$
5. $(A + M) - D$ _____ $B - K$	6. $5 + 2 - a = 7$ _____
$10 = 10$	$20 = 20$
7. $10 - 2 - 2 = 10$ _____	8. 20 _____ $= 20 + b$

obtained for Problems 4 and 5. In the fourth problem it was necessary to preserve the equality by the increase or decrease of both parts by the same value. In the fifth, it was necessary to understand the meaning of a new expression. The problems involving numbers were solved correctly by almost all the children.

The results for similar problems show that many first-grade children mastered well the significance of shifts from equality to inequality and back again. They take into consideration the peculiarities of these shifts when they work with letters as well as when they use numerals and easily complete the required operations. Over the 15 years we taught mathematics using this experimental program, we repeatedly obtained similar results. Thus we conclude, on the basis of the entire complex of tests, that progression through our program insures that children form full and stable conceptions of the basic properties of quantities and operations using them.

CONCLUSION

Our many years of work with this experimental curriculum for the primary grades has led us to several conclusions. First, based on understanding the general characteristics of equalities and inequalities and the shift from one to the other, children's work with numbers can be directed not only to the techniques of calculation, but also to the study of structural relationships that regulate these calculations. In particular, this understanding allows children to form precise ideas about the unity of addition and subtraction (and then multiplication and division), and about the dependence of the change in the results of an operation on the change of its components. Thus operations with numbers can be studied more productively than with traditional methods of teaching.

Second, work with quantities serves as an introduction to numbers—whole numbers as well as fractions. Using the properties of quantities it is possible to decrease the gap between whole numbers and fractions. This, in our view, is an important step when constructing a course of elementary school mathematics. Third, work with quantities is connected from the very beginning with literal symbols. This permits a child to study mathematical relationships themselves, which is very important for further progress in mathematics.

On the basis of our investigations, we believe that an elementary mathematics course should have certain characteristics. A significant portion of time and topics should be directed to introducing a child into the world of concrete objects, which serve as the source of appropriate concepts. The child's development of operations that open up this world permits a later successful transition to more complicated concepts. Therefore, the program should include problems that require the comparison of objects. While solving these problems children learn to isolate the specific relationships of objects being transformed into a quantity. These actions are the starting points for the child's understanding of the meaning of the operations of addition and subtraction and for the mastery of their basic properties.

As a child learns to isolate the relationships of quantities, intermediate methods of representing the results of concrete actions take on great importance. Traditional means of forming concepts do not allow children to represent and model the properties of an object undergoing a dynamic form of action. As the frequency of such modeling actions is reduced, the necessity for intermediate means of description is decreased as well. It is as if the concept itself and the symbolic means of its expression relate directly to the properties of the object. In a school subject, intermediate means of description have crucial significance because they mediate between a property of an object and a concept. Our successful work in the experimental program is due in large measure to the fact that we were able to find and introduce such intermediate means, for example graphic representations of objects and abstract drawings used in isolating and depicting the relationships of compared objects.

Our beginning course in mathematics has yet another feature. By performing operations with manipulatives, the child masters a conception of the general features of a mathematical object. These general features determine the course of further material to be studied. Thus when we single out the area of scalar quantities as the subject for the first section of our course, we have in mind for future study the group of mathematical disciplines clustered around the concepts of the real number and the fundamental mathematical operations.

The general bases for introducing the concept of number as a particular type of quantity have been set forth in this chapter. A complete description of such an introduction can be found in El'konin and Davydov (1962, 1966), and Davydov (1969). The main idea of such an introduction should be reemphasized. Using an understanding of the general properties of quantities a small child can, in particular educational situations requiring intermediate processes for the equalization and comparison of quantities, pick out the multiple relationship of these quantities as that of some whole to one or another of its parts. When children have singled out this relationship, they can progress to the understanding of two important characteristics of numbers. First, an object as a quantity in itself is not determined numerically. It gains a numerical determinacy when a person chooses another quantity as a unit of measurement. Second, one and the same object can be measured with different units and thus be designated with various numbers. If a child in particular problems can freely change the units of measurement of some quantity, designating it with various numbers, then in principle the child is properly oriented toward the origin of the concept of number. That is, the child has mastered the concept proper.

In our view these psychological ideas are foreign to the traditional methods of teaching elementary mathematics. These traditional methods do not develop children's abilities to carry out purposeful mathematical thought. The main purpose of this chapter has been to characterize this kind of thought and show how it can be demonstrated in practice. A further concrete elaboration of these ideas, in our view, will make it possible to improve significantly the practice of teaching mathematics in school.

REFERENCES
[In Russian]

Bourbaki, N. *L'architecture de mathématiques. Les grands courants de la pensée mathématiques*. Paris: Cahiers du Sud, 1948.

Davydov, V. V. (Ed.). [*The mental abilities of elementary school children with respect to mastering mathematics*]. Moscow: Prosveshchenie, 1969.

Davydov, V. V. [*Types of generalization in instruction (Logico-psychological problems in designing curricula*]. Moscow: Pedagogika, 1972.

El'konin, D. V., & Davydov, V. V. (Eds.). [*Questions concerning the psychology of the learning process in elementary school children*]. Moscow: Academy of Pedagogical Sciences of the RSFSR, 1962.

El'konin, D. V., & Davydov, V. V. (Eds.). [*Learning capacity and age level*]. Moscow: Prosveshchenie, 1966.

Kagan, V. F. [*Essays on geometry*]. Moscow: The University of Moscow, 1963.

Kolmogorov, A. N. [Introduction]. In H. Lebesque, *The measurement of quantities* (2nd ed.). (Russian ed.). Moscow: Uchpedgiz, 1960.

Kolmogorov, A. N. [Quantity]. In *The great Soviet encyclopedia* (3rd ed.). Moscow: Sovetskaya Entsiklopediya, 1971.

Lebesque, H. Sur la mesure des grandeurs. *L'enseignement mathématique*. 1933-36, *31-34*.

Author Index

A

Abelson, R. 86, *97*
Allendoerfer, C. B. 60, *66*
Ames, L. B. 100, 102, *114*
Andronov, V. P. 72, *80,* 84, *97*
Anick, C. 1, 2, *6*
Asao, H. 220, *222*

B

Bates, E. 157, *169*
Beckwith, M. 91, *97*
Behrens, M. S. 99, *115*
Beilin, H. 104, *114*
Bekku, S. 219, *222*
Bendix, H. 32, *37*
Berliner, D. 218, *222*
Bessant, H. 29, *37*
Bever, T. G. 104, *115*
Biggs, J. B. 171, 172, 173, *181, 182*
Binet, A. 191, *209*
Binks, M. G. 219, *222*
Bjonerud, C. D. 100, *114*
Blevins, B. 104, 106, 108, 110, 112, *114*
Blume, G. W. 1, 2, *6,* 13, *23,* 64, *66*
Bonnevaux, B. 193, *210*
Bourbaki, N. 229, *238*
Broquist, S. 78, *81,* 101, *115*
Brousseau, G. 58, *59*
Brown, J. S. 117, 120, *135,* 138, 140, *155*
Brown, R. 3, *6*
Brownell, W. A. 1, *6,* 91, *97*
Bruner, J. S. 204, *209*
Brush, L. R. 64, *66,* 104, 106, *114*
Buckingham, B. R. 100, *114*
Burton, R. R. 117, 120, *135,* 138, 140, *155*

C

Carpenter, T. P. 1, 2, *6,* 9, 13, 21, 22, 23, *23, 24,* 57, *59*
Case, R. 157, 159, 160, 161, 162, 164, 165, 166, *169*
Cawley, J. F. 29, *37*
Clapp, F. L. 99, *114*
Cole, M. 193, 198, 199, 204, *209, 210*
Coleman, J. S. 204, *209*
Collis, K. F. 171, 172, 173, *181, 182*
Conne, F. 51, *59*
Cooper, R. G. 104, 106, 108, 110, 112, 113, *114, 115*
Corbitt, M. K. 23, *23*
Cronbach, L. 1, *6*
Cruikshank, W. 29, *37*

D

Dahmus, M. 30, *37*
Dannemiller, J. 113, *114*

Dasen, P. R. 192, *209*
Davydov, V. V. 72, *80,* 84, *97,* 227, 237, *238*
Dixon, R. 32, *37*
Dol, S. 213, *222*
Durand, C. 48, *59*
Dutton, W. H. 100, *114*

E

Eggleston, V. H. 68, *81*
Eggleston, U. G. 91, *97*
Eicholz, R. E. 60, *66*
El'Konin, D. V. 237, *238*
Ezaki, S. 220, *222*

F

Fillmore, C. 32, *37*
Fisher, J. P. 52, 57, *59*
Fleenor, C. R. 60, *66*
Fujinaga, T. 212, 213, 214, *222*
Fukuzawa, S. 218, 220, *222*
Fuson, K. C. 64, 65, *66,* 68, 70, 73, 75, 77, 78, 79, 80, *80, 81*

G

Gallistel, C. R. 64, *66,* 68, *80,* 91, *97,* 100, 102, 104, 113, *114*
Gelman, R. 64, *66,* 68, *80,* 91, *97,* 100, 102, 104, 107, 108, 113, *114, 115,* 163, *170*
Gentner, D. 32, *37*
Gibb, E. G. 9, 10, 13, *24*
Ginbayashi, H. 213, 221, *222, 223*
Ginsburg, H. P. 113, *114,* 191, 193, 199, 200, 204, 207, *209*
Gold, A. P. 166, *170*
Goldberg, J. 162, *169*
Gonzalez, M. 193, *210*
Goodstein, H. A. 29, *37*
Gordon, S. 29, *37*
Goth, P. 104, 106, 108, 109, 110, 111, 112, *114*
Greeno, J. G. 1, *7,* 10, *24,* 32, *37,* 64, *66*
Groen, G. J. 9, 10, 15, 20, 21, *24,* 79, *80,* 101, 113, *114, 116,* 193, 199, *209,* 216, *222*
Grouws, D. A. 10, 11, 22, *24*
Guerry, V. 199, *209*

H

Hall, J. W. 68, 79, *80*
Hashimoto, J. 218, 220, *222*
Haslerud, G. M. 217, *223*
Hatano, G. 212, 214, 219, 220, *222, 223*
Hebbeler, K. 13, *24,* 193, *209*
Helfgott, J. 29, *37*
Heller, J. 1, *7*
Hendenborg, M. 101, *115*
Hetzel, J. 122, *135*
Hiebert, J. 1, 2, *6,* 9, 13, 21, 22, *23, 24*
Hirstein, J. J. 9, *24,* 70, *81,* 84, *97*
Hishitani, S. 219, *222*
Hollingshead, A. 205, *209*
Hosoya, J. 212, 213, 214, *222*
Houlihan, D. M. 193, *209*
Hunt, J. McV. 165, *170,* 204, *209*
Husen, T. 217, *223*

I

Ibarra, C. G. 10, 11, 13, 22, *24,* 65, *66*
Ilg, F. 100, 102, *114*
Imaizumi, N. 212, 213, *223*
Inagaki, K. 212, *223*
Ito, Y. 214, *223*

J

Jensen, A. R. 204, 208, *209*
Jerman, M. 10, *24,* 29, 30, 35, *37*
Johnson, D. C. 33, *38,* 64, *66*

K

Kagan, V. F. 229, *238*
Katriel, T., 10, 13, *24,* 28, 30, 31, 36, *37*
Kaufman, E. L. 91, *97*
Kepner, H. 23, *23*
Khanna, F. 157, *169*
Kiminyo, D. M. 193, *210*
Klahr, D. 3, *6,* 91, *97,* 113, *115*
Knight, F. B. 99, *115*
Kobe City Institute for Education Research, 215, *223*
Kolmogorov, A. N. 228, 229, 230, *238*
Kuhn, T. 2, *6*
Kulm, G. 30, *38*
Kurland, D. M. 161, 162, *169, 170*

L

Labov, W. 204, *210*
Lankford, F. G. 23, *24*

Laurendeau, M. 112, *115*
Lave, J. 193, 194, *210*
Lebesque, H. 228, *238*
Leitner, E. 104, 106, 108, 110, 112, *114*
Leo, T. J. 60, *66*
Lindquist, M. M. 23, *23*
Lindvall, C. M. 10, 11, 13, 22, *24*, 65, *66*
Lingman, L. 101, *115*
Linville, W. 30, 35, *37*
Lipsky, J. 32, *37*
Liu, P. 159, 161, *170*
Loftus, E. F. 10, *24*, 29, *38*
Lord, M. W. 91, *97*
Lorton, P. Jr. 30, *37*
Lyons, J. 32, *37*

M

Marini, Z. 159, 160, *170*
Markman, E. 112, *115*
Marthe, P. 52, *59*
McLaughlin, K. L. 100, *115*
Mehler, J. 104, *115*
Menninger, K. 113, *115*
Mierkiewicz, D. 78, *80*
Mihake, Y. 219, *222*
Milton, K. 60, *66*
Mirman, S. 30, *37*
Mitsuyasu, F. 212, 213, *223*
Mombu-Sho 215, 216, *223*
Moser, J. M. 1, *6*, 9, 13, 21, 22, *23*, *24*, 32, *37*, 57, *59*
Mott, S. M. 100, *115*

N

Nagai, T. 212, *223*
National League of Institutes for Educational Research 211, *223*
Nesher, P. S. 10, 13, *24*, 28, 29, 30, 31, 34, 36, 37
Newell, A. 91, *97*

O

Oakley, C. D. 60, *66*
O'Daffer, P. G. 60, *66*
Ogbu, J. U. 204, *210*
Omanson, S. 149, *155*
Osawa, K. 220, *222*
O'Shea, T. 139, *155*

P

Paige, J. 32, *37*
Parker, T. 193, *210*
Parkman, J. M. 9, 10, 15, *24*, 101, *114*, 216, *222*
Pascual-Leone, J. 160, *170*
Peled, I. 149, *155*
Piaget, J. 109, 111, *115*
Pimm, D. 1, 2, *6*
Pinard, A. 112, *115*
Poll, M. 101, *114*
Popper, K. 25, 26, *37*, 187, *190*
Posner, J. K. 193, 198, 200, 207, *209*, *210*

R

Rees, R. 29, 35, *37*
Reese, T. W. 91, *97*
Resnick, L. B. 9, 10, 20, 21, 22, *24*, 79, *80*, 101, 113, *114*, *116*, 142, 149, *155*, 193, 199, *209*
Restle, F. 91, *97*
Reys, R. E. 23, *23*
Richard, E. E. S. 100, *115*
Richards, J. 186, 189, *190*
Riley, M. S. 1, *7*, 13, *24*
Rosenshine, B. 218, *222*
Rosenthal, D. J. A. 10, 22, *24*
Russell, R. L. 193, 200, 204, 207, *209*

S

Saiga, H. 212, 213, *222*
Samejima, Y. 214, *223*
Saxe, G. B. 113, *115*
Schaeffer, B. 68, *81*, 91, *97*
Schank, R. 32, *37*, 86, *97*
School Mathematics Study Group 60, *66*
Scott, J. L. 68, *81*, 91, *97*
Scribner, S. 193, 198, 199, *209*, *210*
Searle, B. W. 30, *37*
Secada, W. 70, *81*
Sharp, D. 193, *210*
Shikata, J. 218, *223*
Siegler, R. S. 159, 160, 166, *170*
Simon, H. A. 3, *7*, 32, 37, 91, *97*
Skemp, R. R. 183, 186, 189, *190*
Smedslund, J. 104, 106, 107, 110, 112, *115*
Spelke, E. 113, *115*
Spikes, W. C. 9, *24*, 70, *81*, 84, *97*
Starkey, P. 102, 104, 106, 108, 110, 111, 112, 113, *114*, *115*

Steffe, L. P. 9, *24*, 33, *38*, 64, *66*, 70, *81*, 84, *97*, 186, 189, *190*
Stevenson, H. W. 193, *210*
Suga, Y. 214, 220, *222*
Suppes, P. 1, *6*, 9, 10, 15, *24*, 29, 30, *37*
Svenson, O. 78, *81*, 101, *115*

T

Tatsuno, C. 218, 220, *222*
Teubal, E. 30, 34, *37*
Thibodeau, G. 29, *37*
Thorndike, E. 1, *7*
Thyne, J. M. 102, *115*
Toyama, H. 219, 221, *223*

U

Umegaki, M. 219, *223*
Umemoto, T. 217, *223*

V

Van Lehn, K. 124, 126, *135*
Vergnaud, G. 32, *38*, 48, *59*
Vitello, S. 29, *37*
Vlakokos, I. 29, *37*
Volkman, J. 91, *97*
von Glasersfeld, E. 83, 91, *97*, 186, 189, *190*

W

Wallace, J. G. 3, *6*, 91, *97*, 113, *115*
Watson, R. 161, *169*
Wheat, M. 30, *38*
Wheeler, L. R. 99, *115*
Wilkinson, A. 193, *210*
Williams, A. H. 100, *116*
Winston, P. H. 134, *135*
Woods, S. S. 9, 20, 21, *24*, 101, *116*
Woody, C. 100, *116*

Y

Yamauchi, M. 219, *222*
Yoshimura, T. 212, 216, *223*
Young, R. M. 139, *155*

Z

Zaslavsky, C. 113, *116*, 192, 193, *210*

Subject Index

A

Abacus, 151, 215, 219
Abstract unit items, 77, 83
Accommodation, 3
Adding-on, 17, 72, 73
Addition, written, 192
Additive structure, 40, 58
Algorithm(s), 120, 136
  buggy, 138
  computational, 201, 217
  subtraction, 1
Alphabet substitution for counting, 75, 78
Assimilation, 3
Associative, 105, 229, 231, 235

B

Baoulé, 198
Borrow, 118, 138, 147
"Bug(s)", 117, 138
  compound, 120
  primitive, 120
BUGGY, 117

C

Capacity, 176
Canonical (displays), 143
Cardinal, 68-73, 101
  principle, 68, 69
Cardinality rule, 68
Carry, 147
Central processing-capacity, 160, 167
Chisenbop, 75
Chunking, 160
Class, social, 191
Class inclusion, 52
Clinical interview, 5, 13, 189
Closure, 177, 179
Cognitive,
  development, 156, 171
  processes, 3
  structure, 171
  system, 191
    cultural, 192
    formal, 192
    informal, 191, 203
    natural, 191
Commutative, 62, 105, 114, 131, 229, 231, 235
Comparison, 10, 17, 32
Compensation, 105-109, 110
Composition, 42
Concrete operations, 112, 114
Conceptual,
  field, 40, 58
  structures, 184
Conservation of number, 57, 110-113, 212, 214

Contextual constituent, 30
Continuous, 63, 226
Coordination, bifocal, 158, 160, 163
Correspondence, one-to-one, 191
Count(ing), 100, 168
  -all, 14, 67, 68, 69, 71, 72, 76, 77
  backward, 20, 100
  double, 75
  -down, 17, 221
  -on (up), 15, 17, 21, 67-81, 100, 167, 173, 193, 212
    actual, 72
    formal, 72
    keeping track methods, 73-77
  recursive, 100
  sequences, 14
  strategies, 9, 22
  type, 83
Counters,
  with figural unit items, 84, 89-91, 96
  with motor unit items, 85, 91-96
  with perceptual unit items, 84, 86-89, 91, 96
Culture, 5, 191, 211, 217

D

DEBUGGY, 117
Derived number facts, 79
Diagnostic system, 117
Dioula, 194
Director system, 184
Discrete, 63, 226
Double counting, 75
Dynamic, 32, 46, 52

E

Elaboration, 158, 163
Enrichment, 152
Estimating, 14
Equalizing, 10, 233
Extended abstract (stage), 176, 180
Extension, 69

F

Figural,
  representation, 71
  unit items, 77, 83, 84, 89-91, 96
Function, 60, 229

G

Goal state, 184

H

Hill climbing, 91
Horizontal décalage, 112

I

Information processing, 3
Intelligence, 184
Interviews, 13, 142, 189
Intuitive extension, 93-96
Invention, 142
Inversion, 105-109, 110, 113
Ivory Coast, 194

L

Learning, 184
Lexical constituent, 30
Linguistic, 219
Logical structure, 28, 36

M

Matching, 18, 21
Mapping, 147
  code, 147
  operations, 153
  result, 147
Missing addend problem, 28, 91, 94-96
Motor unit items, 77, 83, 85, 91-96
Multiplicative structure, 40
Multistructural, 176, 178

N

Normal science, 2, 6
Notation,
  conventional, 65
  mediating, 65
  symbolic, 65
Numerical calculus, 40
Numerically quantified arguments, 28

O

Object permanence, 85
One-to-one correspondence, 191